Routledge Contemporary Human Geography Series

Urban Geography 3rd edition

Tim Hall

LONDON AND NEW YORK

First published 1998 by Routledge

Reprinted 1998

Second edition published 2001 by Routledge

Reprinted 2002, 2005

Third edition published 2006 by Routledge
2 Park Square, Milton Park, Abingdon, Oxon OX14 4RN

Simultaneously published in the USA and Canada
by Routledge
270 Madison Ave, New York, NY 10016

Routledge is an imprint of the Taylor & Francis Group

Typeset in Times and Franklin Gothic by
Keystroke, Jacaranda Lodge, Wolverhampton
Printed and bound in Great Britain by
MPG Books Ltd, Bodmin

British Library Cataloguing in Publication Data
A catalogue record for this book is available from the British Library

Library of Congress Cataloging in Publication Data
Hall, Tim, 1968–
Urban geography / Tim Hall.— 3rd ed.
p. cm. — (Routledge contemporary human geography series)
Includes bibliographical references and index.
1. Urban geography. I. Title. II. Series.
GF125.H35 2006
307.76—dc22 2005020301

ISBN10: 0–415–34445–X ISBN13: 9–78–0–415–34445–6 (hbk)
ISBN10: 0–415–34446–8 ISBN13: 9–78–0–415–34446–3 (pbk)

Urban Geography
3rd edition

Cities across the world are changing rapidly, influenced by the forces of economic and cultural globalisation and by the pressures for environmental, economic and social sustainability. City economies, landscapes, images, environments and social geographies are changing as a result of these forces.

This third edition of *Urban Geography* continues to examine the new geographical patterns that are forming within cities and the ways in which geographers have sought to make sense of this urban transformation. Introducing both traditional and contemporary approaches and perspectives in urban geography, this book examines the globalisation of the urbanisation process and explores ways in which governments and institutions have responded to the resulting challenges.

This extensively revised new edition, styled in a student-friendly chapter format, includes a new chapter on urban regeneration and comprehensively updated case studies, all illustrated by numerous figures, photos and tables. This is a valuable introduction to urban geography for all geography students.

Tim Hall is Lecturer in Human Geography at the University of Gloucestershire.

Routledge Contemporary Human Geography Series

Series Editors:
David Bell, Manchester Metropolitan University
Stephen Wynn Williams, Staffordshire University

This series of texts offers stimulating introductions to the core subdisciplines of human geography. Building between 'traditional' approaches to subdisciplinary studies and contemporary treatments of these same issues, these concise introductions respond particularly to the new demands of modular courses. Uniformly designed, with a focus on student-friendly features, these books will form a coherent series which is up-to-date and reliable.

Existing Titles:

Cultural Geography

Tourism Geography

Development Geography

Political Geography

Geographies of Globalization

Urban Geography, 3rd edition

For Cathy

Contents

Plates

Figures

Tables

Case studies

Acknowledgements

I would like to thank all of those colleagues and students at the University of Gloucestershire who have contributed to the development of this book over a number of years. Various bits of this book have cropped up in, and been refined through, invaluable discussions both in class and outside. I would also like to thank everyone at Routledge who has helped keep all three editions of this book on track. Finally, thanks to Cathy for her help and contribution, again to all three editions of the book.

The author and publishers would like to thank the following for granting permission to reproduce images in this work:

Plates
John Rennie Short for supplying the Syracuse plates.

Matt Halstead, for a photograph of Graffiti in Barcelona.

Figures
Elsevier, for the figure 'Mann's model of the British city' from Knowles and Wareing, *Economic and Social Geography Made Simple*, 1976.

Routledge, for the figure 'The post-industrial global metropolis', from Graham and Marvin, *Telecommunications in the City*, 1996.

The Geographical Association, for the figure 'A fringe-belt model' from Whitehand, 'Development cycles and urban landscapes', *Geography*, 71 (1), p. 11, 1994.

Tables
Blackwell Publishing Ltd, for the table 'Global cities and corporate headquarters', from Johnston *et al.* (eds), *Geographies of Global Change*, 1995.

Sage Publications, for the table 'The evolution of urban regeneration', from Roberts and Sykes, *Urban Regeneration: A Handbook*, 1999.

John Wiley & Sons Ltd, for the table 'Local authority promotional packages, 1977 to 1992', from Gold and Ward, *Place Promotion*, 1994.

Every effort has been made to contact copyright holders for their permission to reprint material in this book. The publisher would be grateful to hear from any copyright holder who is not here acknowledged and will undertake to rectify any errors or omissions in future editions of this book.

1 Why urban geography?

It should be apparent to you, before you even start to study the subject, that urban geography is more than just another subject on a university or college curriculum. Many of the most pressing issues facing contemporary societies now and into the future revolve around life in cities. Life in cities is replete with both problems and possibilities. Cities represent both the best and worst of life in the earliest twenty-first century. If you wish to engage with the world outside your window, either in theory, in practice or through your everyday life, it is almost impossible to do so without at least some engagement with urban geography.

If you need convincing of the importance of urban geography today, think for a second about what are the most pressing issues in your life. Among specific concerns about family it is likely that they will include:

- Where you live
- Who you live among
- Your opportunities for leisure and social activities
- Your personal mobility
- Your income, career opportunities and access to wealth
- Your personal safety and your exposure to antisocial behaviour
- Your health and levels of stress
- Your access to facilities such as financial and health services
- The pollution of your local environment

These are all issues at the heart of urban geography and now touch the lives of the majority of the world's population. Many of them are examined in this book.

Perhaps the most pressing reason, though, why urban geography is key to social, economic and environmental sustainability in the future concerns the huge inequalities that are evident in the cities of the world. Inequality is a theme that runs throughout this book and crops up, in one form or another, in almost all of its chapters. The gaps between the richest and the poorest in society have been growing throughout the world over the past twenty years or so. However, while income levels are perhaps the key determinant to life chances, they do not tell the whole story of this inequality. The poorest in society suffer the worst housing, if indeed they can find housing at all, the most dangerous and polluted environments, the shortest life expectancy, the most restricted mobility and the worst access to services. Contrasts between the life chances of individuals are found at their most stark in many urban areas where, for example, islands of wealth and redevelopment abut the most run-down neighbourhoods and threaten their existence.

Understanding the processes that produce and sustain these inequalities and imagining possible solutions necessitates engagement with life in cities and hence with urban geography.

2 New cities, new urban geographies

Five key ideas

- **Cities are in a constant state of change.**
- **At times this change is fundamental and transforms the nature of cities.**
- **Contemporary cities are extremely diverse in their nature.**
- **A number of urban models have been devised that describe the general characteristics of different types of cities.**
- **Approaches to urban geography have evolved and changed through time.**

New cities, new urban geographies?

The only consistent thing about cities is that they are always changing. Classifying and understanding the processes of urban change present problems for geographers and others studying the city. Cities, since their inception, have always demonstrated gradual, piecemeal change through processes of accretion, addition or demolition. This type of change may be regarded as largely cosmetic and the underlying processes of urbanisation and the overall structure of the city remain largely unaltered. However, at certain periods fundamentally different processes of urbanisation have emerged; the result has been that the rate of urban change has accelerated and new, distinctly different, urban forms have developed. This occurred, for example, with the urbanisation associated with industrialisation in the UK in the nineteenth century.

Geographers have constantly to ask themselves whether the changes they observe are part of the continual process of piecemeal change or whether they are part of more fundamental processes of transformation. Just such a debate occupied geographers, sociologists and other social scientists in the latter part of the 1980s and the early 1990s. The issue of whether we are witnessing the emergence of new types of cities has also raised questions about the adequacy and relevance of the geographical models and theories developed in the past to understand cities.

The earlier mention of the Industrial Revolution raises issues of investigation that shape the themes of this book. Do we need to look at the changes in not only the national but also the international economy since the 1970s and ask ourselves whether or not they are as epochal in their extent and significance as those changes now labelled the 'Industrial Revolution'? The answer to this question is unequivocally yes. There is little doubt that since the early 1970s the world economy has been affected by a number of fundamental changes. The ramifications of these changes have been enormous and have affected not only the economic life, but also the social, cultural and political lives of nations, regions, communities and individuals. Tracing the links between the changes in the world economy and those in the landscapes, societies, economies, cultures and politics of cities is one aim of this book.

Visually, the evidence of a fundamental transformation of the processes of urbanisation appears compelling. The signs of significant change are apparent in many urban landscapes of North America, the UK, mainland Europe and many parts of the developing world. Some of the most widely debated of these signs of change have been the enhancement of city centres by extensive redevelopment, the redevelopment of derelict, formerly industrial areas such as factories and docks, the use of industrial and architectural heritage in new commercial and residential developments, the social, economic and environmental upgrading of inner-city neighbourhoods by young, middle-class professionals (a process referred to as 'gentrification'), the appearance of brand-new 'city-like' settlements on the edges of existing urban areas, and the emergence of large areas of poverty and degradation (often referred to as social exclusion), for example, in old inner-city areas, and on council housing estates on the edges of numerous towns and cities.

The language that academics have used to debate and describe contemporary urban change would suggest that some profound differences in the urbanisation process have emerged. The language used by academics to describe these changes has included: from industrial to post-industrial, from modern to postmodern, from Fordist to post-Fordist. However, despite apparently compelling visual evidence and the language used to describe change, it is important to try to remain objective and to assess the degree to which these changes could truly be called a transformation of the urbanisation process and an emergence of new forms of urban settlement.

Some of the questions that emerge from this debate include:

- How significantly has urban form been altered?
- How have these changes varied geographically between different cities?
- How differently does urban life feel now, and for whom?

This book explores these, and other, questions. It will consider the structural, economic, political, social and cultural changes that have affected urbanisation, primarily in the UK, but with reference to mainland Europe and North America.

Discussion topic

Think of a city that you know well. What evidence can you find that processes of urban change are taking place there? This evidence may include observations of the urban landscape, planning documents or development proposals, media reports or academic writing. Does the evidence you have collected suggest that this is a profound, epochal change or more ongoing piecemeal change, or some combination of the two?

Different types of city

Many urban geographers and historians have argued that the cities we recognise are the product of a long evolutionary process, during which the settlements of 15,000 BC gradually evolved into the complex cities of the early twenty-first century. This view may seem very appealing; however, it ignores some very important dimensions of contemporary urbanisation. No two cities are identical. They may be broadly similar, but cities have very different landscapes, economies, cultures and societies. This is a reflection of the fact that cities are shaped by a diverse set of processes. The particular set of processes that affect city development depends on a number of factors that are unique to individual cities, such as city size and the nature of its economy, and/or related to wider factors, such as the relationships between networks of cities, the nature of the nation within which they are located and their position

Plate 2.1 ***Postmodern urbanisation, the Olympic Port, Barcelona***

within the world economy. The diversity of city types and processes of urbanisation cannot be reduced to a simple, linear evolutionary process. It is preferable to adopt a perspective that recognises this diversity and to think of cities as having different roles and positions in the world economy. The trajectory of urban development is bound up with the workings of the world economy and the relationships of individual cities to this (Savage and Warde 1993: 38). The following classification of different types of cities recognises this:

- Third World cities
- Cities in socialist countries
- Global (world) cities
- Older (former) industrial cities
- New industrial districts

(Savage and Warde 1993: 39–40)

This classification is not totally comprehensive, nor should it be applied too rigidly. For example, many cities fall into more than one of the categories listed (Savage and Warde 1993: 40). London is a global city; however, it contains a considerable decaying industrial economy and yet it is surrounded by many new industrial districts. It is often far from easy to determine which category describes a city best. Further, there is a great deal of diversity within each category, especially between Third World cities but even between older industrial cities which may have been based upon different industries. Despite these limitations this classification recognises that urbanisation is different in various parts of the world and for different types of city (Savage and Warde 1993: 38–40). This book focuses primarily on the urban geography of older industrial cities, global cities and new industrial spaces.

Discussion topic

Using the example of the city you thought of for the previous discussion topic, which of the categories outlined by Savage and Warde above do you think it fits into? Can you justify your answer, and what evidence did you use to come to this conclusion? Does it fit neatly into one category or does it demonstrate some characteristics of other categories? What does this suggest to you about the problems of classifying urban areas?

Evolution of the industrial city

Cities which were formed, or in large part influenced, by the processes of urbanisation linked to the 'industrial revolution' of the nineteenth century are of relevance to a discussion of contemporary urban geography for a number of reasons. First, such 'industrial' cities constitute a large part of the urban systems of the UK, the USA and Europe. Industrialisation influenced the internal geographies of many cities in these regions as well as the economic, political and physical links between them. These legacies have formed important dimensions of subsequent urbanisation. These cities are variously referred to in debate as 'modern' or 'industrial'.

As well as the obvious importance of these cities to the urban geography of older industrial nations, the industrial city has also had a disproportionate influence on modern urban theory. The distinctive built form of the industrial city became well known in the latter half of the nineteenth century. The dreadful conditions that prevailed in its inner residential areas made it ripe for study by journalists, satirists, social reporters and novelists. Novels which explored the dark sides of these cities include *Sybil: The Two Nations* by Disraeli (1845), *North and South* by Elizabeth Gaskell (1848) and *Hard Times* by Charles Dickens (1854); social reports included Frederick Engels' exploration of Manchester published in *The Condition of the Working Class in England* (1844) and Charles Booth's social survey of London published as *Life and Labour of the People of London* (1889). The horrific images of these texts did much to form the initial middle-class reaction to the industrial city and the lasting perceptions still present today.

In North America the link between urban change and the formation of urban theory was even more direct. For much of the early twentieth century the most influential sociology department in North America was at the newly formed University of Chicago. The specialism of this department was urban sociology and it was in Chicago that much of their research was conducted (Ley 1983: 22). Some of the most famous publications of this school included Robert Park, Ernest Burgess and R.D. McKenzie's *The City* (1925), Ernest Burgess' *The Urban Community* (1926), H. Zorbaugh's *The Gold Coast and the Slum* (1929) and H. Hoyt's *One Hundred Years of Land Values in Chicago* (1933). Chicago was a new city at the time; it had grown rapidly and owed much of this growth to industrialisation. Models of urban structure, most famously Burgess' concentric zone model (Figure 2.1) and Hoyt's sector model (Figure 2.2), were based on this research, and inevitably reflected the structure of the city and the forces that created it. The influence of these models on Mann's (1965) model of the British city is obvious (Figure 2.3).

Reactions to the industrial city also shaped a number of nineteenth- and twentieth-century traditions in the fields of political ideology, town planning and literature (Short 1984: 14–16). The physical and intellectual legacy of the industrial city, therefore, is hard to ignore.

The most immediately striking aspects of the industrial city were the extent to which they revolutionised the urban systems of the UK and the USA and the speed at which this took place. In Britain in 1800, for example, only one city, London, exceeded 100,000 people; however, by 1891 this had risen to twenty-four cities (Ley 1983: 20). In the USA this process occurred some fifty to sixty years later. Its effects were none the less fundamental. In both the UK and USA the focus of their urban

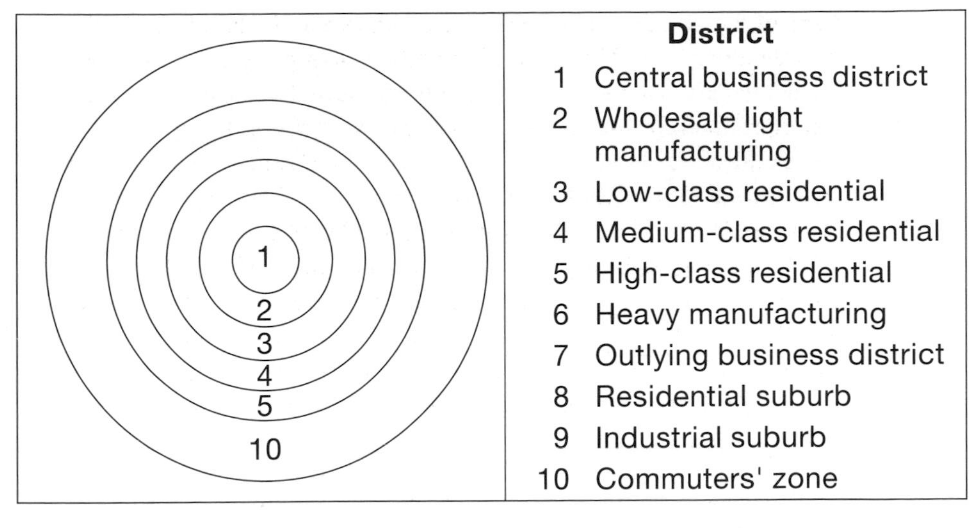

Figure 2.1 ***Burgess' concentric zone model***

Source: Ley (1983: 73)

District

1 Central business district
2 Wholesale light manufacturing
3 Low-class residential
4 Medium-class residential
5 High-class residential
6 Heavy manufacturing
7 Outlying business district
8 Residential suburb
9 Industrial suburb
10 Commuters' zone

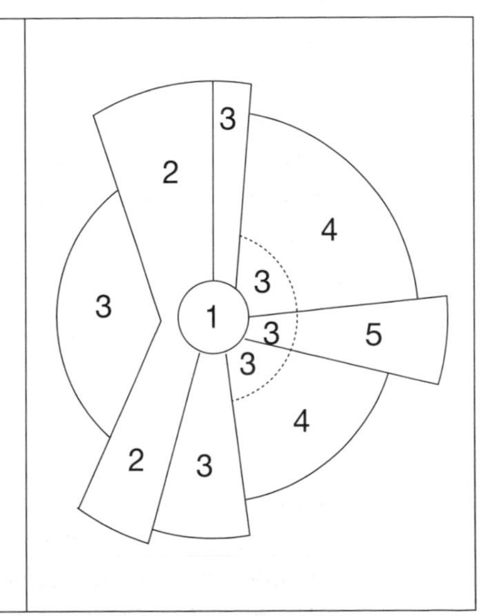

Figure 2.2 ***Hoyt's sector model***

Source: Ley (1983: 73)

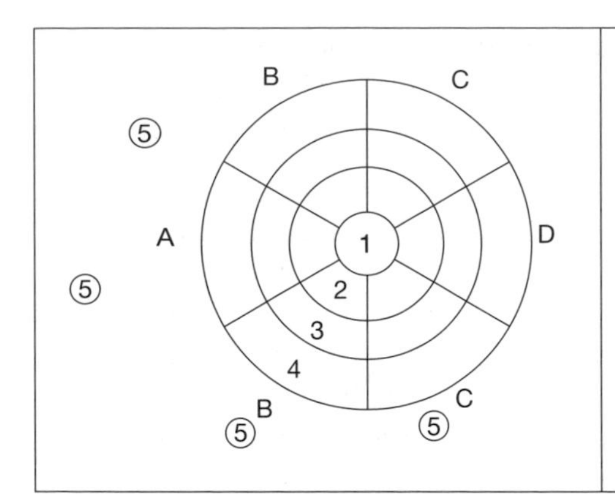

A Middle-class sector
B Lower middle-class sector
C Working-class sector (and main sector of council estates)
D Industry and lowest working-class sector

1 CBD
2 Transitional zone
3 Zone of small terraced houses in Sectors C and D; larger by-law housing in Sector B; large old houses in Sector A
4 Post-1918 residential areas, with post-1945 housing on the periphery
5 Commuting-distance 'dormitory' towns

Figure 2.3 ***Mann's model of a British city***

Source: Knowles and Wareing (1976: 245)

systems shifted away from previous political, religious or mercantile centres towards manufacturing centres. Sources of cheap energy, first water and later coal, acted as strong magnets for this growth, coupled with the accessibility offered by rivers and other waterways such as lakes.

Not only were the size and rapidity of growth of these cities new but so also were the forces that shaped this growth. The growth of the industrial city was tied to the development of the factory system within the emergent form of industrial capitalism. The ability of industry to outbid other uses for land near the centres of cities was fundamental to the generalised form of the industrial city portrayed in contemporary accounts and urban models. The cores of industrial cities remained largely commercial, a land use that was able to outbid all others for this expensive accessible land. However, surrounding this core was typically a ring of industry, which required a large labour force housed nearby, leading to the development of a ring of working-class housing surrounding it. The regulation of this residential development was non-existent until the late nineteenth century in many industrial cities. Systems of urban government were archaic, based on the parish system and more suited to the administration of rural areas or small urban areas. Consequently, they were totally overwhelmed by the scale of urbanisation during the nineteenth century. The quality of the housing in this zone was extremely poor and the provision of services and utilities such as running water, lighting and sanitation was frequently nil. These zones became notorious for outbreaks of lawlessness and disease, initiating a series of 'moral panics' among the wealthier middle classes. These reactions to the zones of new working-class housing were fundamental to the desire of the middle classes to move away from the centre of cities into the expanding suburbs, a desire facilitated by progressive transport innovation.

Despite the theoretical importance afforded to both Manchester and Chicago, they were not typical of the industrial city. Rather, these two cities were the 'shock' cities of industrialisation. They represented extremes where an industrial urban form became most developed and complete. In other cities industrialisation was mediated by local conditions, the legacy, be it physical or otherwise, of earlier rounds of urbanisation. In many cases the transformation of urban form by industrialisation was partial rather than total, mixed in with earlier urban forms, or the form of industrialisation differed from the norm. In Birmingham, for example, initial mass industrialisation was concentrated in small workshops rather than factories. It was not until the early twentieth century that the factory system came to dominate. Further, Birmingham's urban form was disrupted by a complex patchwork of landownership patterns which determined land uses as much as the forces of industrialisation did.

Los Angeles and models of the post-industrial global metropolis

> Los Angeles has been a particularly vivid context from which to explore postmodern urbanization in virtually all its dimensions. I have called it the quintessentially postmodern metropolis not because I see it as a 'model' for all other cities to follow or as a doomsday scenario to warn the rest of the world. If there is anything which places Los Angeles in a special position with respect to an understanding of postmodern urbanization processes, it is that comprehensive vividness I referred to, the particular clarity these restructuring processes have taken in this region of Southern California. In part, this is due to the relative absence of residual landscapes derived from preindustrial, mercantile, and nineteenth-century industrial urbanizations. Although founded in 1781, Los Angeles is pre-eminently a twentieth century metropolis. Since 1900, the population of the Los Angeles region has grown by more than 14 million, more than almost any other city in the world.
>
> (Soja 1995: 128)

In the late twentieth century Los Angeles assumed a position with regard to urban theory comparable to that of Chicago in the early twentieth century. A number of academic and popular accounts have constructed it as an archetype of contemporary and future urbanisation. The shift in the location of the most influential North American school of urban theory symbolises the eclipse of the USA's 'rust-belt' of manufacturing cities by emerging cities of the west coast oriented more to the developing world than to Europe (Savage and Warde 1993: 58). Just as Chicago was home to an influential school of urban sociology, Los Angeles has been home to an influential school of urban studies, the Graduate School of Architecture and Urban Planning at University College Los Angeles, since the early 1970s. This school has included a number of prominent urban theorists who have used Los Angeles as the primary laboratory in which they have researched the emergent processes of urbanisation of the late twentieth century. Important works from this 'California School' have included Allen Scott's *Metropolis: From the Division of Labour to Urban Form* (1988), Ed Soja's *Postmodern Geographies: The Reassertion of Space in Critical Social Theory* (1989) and *Thirdspace: Journeys to Los Angeles and Other Real and Imagined Places* (1996), Mike Davis' *City of Quartz: Excavating the Future in Los Angeles* (1990) and Frederick Jameson's *Postmodernism, or the Cultural Logic of Late Capitalism* (1992). More popular accounts that have explored notions of new processes of urbanisation in cities such as Los Angeles, San Francisco, Tokyo and, to a lesser extent, London have included Deyan Sudjic's *The 100 Mile City* (1993). To these may be added a host of articles in academic and professional journals, magazines and newspapers by the authors listed above and many more.

A major theme of many of these works has been the idea of the fragmentation of urban form and its associated economic and social geographies, namely that the city is ceasing to exist as a recognisable single, coherent entity. Rather it is physically fragmenting as independent cities emerge on the edge of existing metropolises, and economically, socially and culturally fragmenting as divisions between different social groups widen to the extent of their becoming broken. According to this logic, the city fragments into a series of independent settlements, economies, societies and cultures. This is expressed in the idea of the galactic metropolis proposed by Peirce Lewis in 1983 which describes urban form as resembling a series of stars floating in space, rather than a unitary, coherent entity with a definable centre (Knox 1993). This idea of fragmentation was present in two models proposed of the post-industrial city based on research in Los Angeles and Atlanta. These models are the urban realms model (Hartshorn and Muller 1989) which describes a

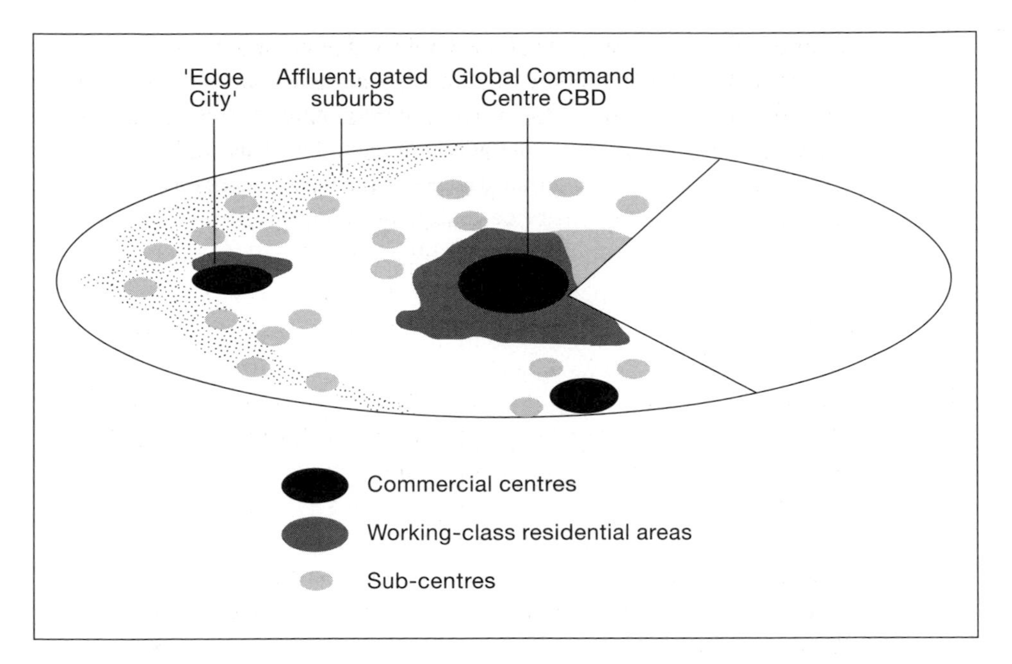

Figure 2.4 ***The post-industrial 'global' metropolis***

Source: Graham and Marvin (1996: 334)

series of separate cities existing within a larger metropolitan area, and Soja's (1989) model of the post-industrial 'global' metropolis (Figure 2.4).

Much of the work of the California School is based on what they argue is a link between changes in the organisation of capitalism (regime of accumulation), and the new industrial sectors and spaces that they throw up, and urban form. These commentators have painted a picture of Los Angeles as a city whose economic and social geographies are based upon new economic growth sectors such as animation (Christopherson and Storper 1986), the motion picture industry (Scott 1988) and hi-tech defence-related industry as well as informal, quasi-legal or illegal economic activities of various kinds (Soja 1995).

However, such models may be subject to two sets of criticisms. The first concerns their attempts to link changes in the regime of accumulation with the restructuring of urban space and neighbourhood formation through the medium of industrial restructuring. To use this as an explanatory basis for neighbourhood formation is very tenuous and narrow, and ignores a number of very important issues in neighbourhood formation and reproduction. This reduction to economic causality reflects a number of other weaknesses in this approach. It pays very little attention to the micro-social struggles and conflicts that take place within and between neighbourhoods. This reflects a general neglect of human action and a focus primarily on the abstract and the structural. Even within the economic sphere their focus is very narrow, since it fails to pay much attention to the service sector, an important component of any advanced urban economy (Savage and Warde 1993: 59–61).

The second set of criticisms are similar to those levelled at models derived from research in Chicago in the early twentieth century. Put simply, how typical is Los Angeles of cities in the late twentieth century? Earlier in this chapter it was pointed out that cities have a diversity of contrasting histories. It should be immediately apparent that to suggest that Los Angeles can provide anything like a general model of urbanisation is flawed thinking. It would be naive to suggest that California School commentators do not recognise Los Angeles' exceptional history. However, as much as a result of the prominence attached to their work on Los Angeles, the city has assumed a theoretical primacy within contemporary debates on urban form. Much of what follows in this book examines the question, implicitly at least, of the extent to which 'industrial' urban forms have been superseded by 'post-industrial' forms. The question of whether Los Angeles represents a common urban future is one for the reader to actively address.

The most profitable application of the work of the California School of urban geography lies not in the issue of modelling overall urban forms, but rather the recognition they afford to emergent trends in processes shaping urban landscapes, economies and cultures, which they discerned first, and with most clarity, in Los Angeles. These trends include the rise of new flexible forms of economic organisation and production; the increasing influence of an interconnected global economy and the crucial role played by cities within this; the appearance of 'edge-cities'; the heightening of various forms of economic, social and cultural inequality within cities which are expressed in new patterns of spatial segregation; the rise of 'paranoid' or carceral architecture based on protection, surveillance and exclusion; and finally the increasing presence of simulation within urban landscapes, imaginations of alternative cities to the 'dreadful reality' of actual cities (for example, theme parks, themed, policed shopping malls), and more subtle forms of simulation that invade everyday life. Ed Soja summarised these trends as 'six geographies of restructuring' (Soja 1995: 129–37). While these should not be uncritically imported as a blueprint for post-industrial urban change, they signal processes discernible to varying extents in many urban localities.

New cities?

Clearly cities, and the processes that produce them, have changed and will continue to change. However, it is important to try to understand the significance of these changes. Are we witnessing the emergence of new types of cities produced by new forms of urbanisation, or are cities remaining fundamentally unaltered except for some cosmetic changes? Or is the situation somewhere between the two?

On the transformation side it is clear that the changing nature of international capitalism (which is explored in more detail below) has had a major impact on the relations between cities and the internal geography of the city. However, while an undoubted resorting has occurred, the overall structure of the city is still recognisably modern rather than postmodern. The structure of the city that has evolved over the course of the twentieth century and been structured by industrial capitalism and national systems of town planning has far from been completely overwritten since the early 1970s.

Rather than a new city emerging, it would be truer to say that 'newer' formations have begun to appear within a more traditional structure. The extent to which the newer elements will come to dominate is difficult to determine. The city is currently experiencing more than just piecemeal change; however, it is some way short of being completely transformed (Cooke 1990: 341; Soja 1995: 126).

A framework for examining urban change

The urban landscape does not simply appear overnight, rather like a movie-lot springing up on vacant land, but has to be *produced*. A large number of actors are involved in this production: architects, designers, builders, property developers and construction workers among others. As well as those directly engaged in the physical production of the urban landscape there are a similarly large number less directly engaged but none the less just as vital. These include investors in the built environment. The composition and organisation of these groups has changed significantly since the mid-1980s. It has also come to include new and increasingly influential players such as pension and finance companies (Knox 1991, 1993).

The technology employed in the production of the built environment has always profoundly affected its form. The development of innovative architectural forms, such as the skyscraper in the early twentieth century, was the product of the coalition of a number of important factors, not least the development of both building and communications technology. The transformation of the technologies of building production (for example, with the employment of computer technology) has profoundly affected the production of the built environment and its form. Each of these components of the production of the built environment is examined throughout the book (Knox 1993).

The urban landscape is not produced without constraint. It cannot simply be thrown up at the whim of a developer or architect. Rather, the built environment is subject to strict control by the state through the planning system. Not only is the urban landscape produced but it is also *regulated*. The regulatory action of the state is an important influence on the development and the direction of physical change in the urban landscape (Greed 1993).

The urban landscape is not produced only to be looked at, although its appearance is a vital component; it is built primarily to be used. The urban landscape serves a vast variety of sectors, be they residential, commercial, industrial, retail or leisure use. Not only is the city produced and regulated but it is also *consumed*. The composition of these groups of consumers and their needs, wants, tastes and ability to consume will profoundly affect what is built for them. Understanding the relationships that surround the changing urban landscape requires that this dimension is also taken into account (Harvey 1989a: 77). Again the consumption dynamics surrounding the urban environment have displayed important changes since the mid-1970s.

The framework outlined above has so far concentrated on the physical components of the city, the production, regulation and consumption of the built environment. However, this is not the sole concern of this book; rather it provides both a fulcrum and a point of departure from which to examine many more aspects of urban life. The early part of the book focuses on factors affecting the production and regulation of urban change, landscape, society and the economy. These are brought together in Chapters 6 and 7. While the focus in these chapters is on changing urban landscapes and images of cities, they also highlight the ways in which these are structured within certain germane economic, political, social and cultural contexts. Chapters 8 and 9 consider how these changes, in turn, structure new urban social, cultural and environmental geographies. In doing so they consider the economic, political, social and cultural impacts of recent urban change.

New urban geographies?

Geography and related disciplines have a long history of models and theories formulated to understand the city and the processes of urbanisation. These have stemmed from several, often contradictory perspectives. Time has witnessed the continual replacement of successive schools of thought with alternatives.

The inadequacy of individual urban models and theories has been continually exposed. Either they have been shown to have been theoretically weak or partial, or that cities have changed to such a degree that the models have rapidly become redundant. The question of the adequacy of previous models of the city is again relevant, given the transitory nature of contemporary urbanisation. Chapter 3 reviews briefly the history of urban theory in the twentieth century and considers the question: Are existing ways of understanding the city becoming redundant?

Project idea

Who is involved in the production of the urban landscape? Collect evidence for a town or city you know well from all of those you feel are involved at a variety of scales. Where do these agents come from, are they local, national or international? These might include builders, architects, developers and investors among others. What constraints or regulations are they subject to and who is responsible for these regulations?

Essay titles

- 'Urban models hide more than they reveal.' Discuss.
- To what extent are the emergent processes of urbanisation recognised by the Los Angeles school geographers apparent in cities internationally?

Further reading

Fyfe, N. and Kenny, J. (eds) (2006) *The Urban Geography Reader*, London: Routledge. [Key reference work. Companion volume to *The City Reader* (see below).]

Knox, P. and Pinch, S. (2000) *Urban Social Geography: An Introduction*, Harlow: Prentice-Hall (ch. 2). [Good introduction to the evolution and development of cities.]

LeGates, R.T. and Stout, F. (2003) *The City Reader*, London: Routledge (2nd edn). [Key reference work containing many readings of relevance to this chapter.]

Mumford, L. (1991) *The History of the City*, Harmondsworth: Penguin. [A classic history of the city.]

Scott, A. J. and Soja, E.W. (eds) (1998) *The City: Los Angeles and Urban Theory at the End of the Twentieth Century*, Los Angeles, CA: UCLA Press. [Key introduction to the work of the Los Angeles School urban geographers and others working in the city.]

Web resources

Centre for Advanced Spatial Analysis, University College London – www.casa.ucl.ac.uk

Chicago Past and Present – www.historyillinois.org/hist.html#chicago

Los Angeles City Website – www.ci.la.ca.us

3 Changing approaches in urban geography

Five key ideas

- **Definitions of 'urban' vary internationally.**
- **Urban areas are not subject to a unique set of processes or influences.**
- **Urban geography has been subject to a series of radical changes of approach throughout its history.**
- **Urban geography has been concerned with the description, interpretation and explanation of urban change and development.**
- **Recently urban geography has been characterised by plurality and a wariness of the explanatory power of any single perspective.**

What is urban?

What is urban? While this may seem a good place to start, it may also appear to be a stupid question. Most people know an urban area when they see one. However, the point of the question is not recognition but definition. How are we able to define 'urban'? What, if anything, is uniquely urban? What makes urban areas qualitatively different from other types of areas?

It is relatively easy to classify which areas are urban and which areas are not urban. There are a host of indicators available. These include population size, population density, number and range of services and employment profiles. Some sociologists have even claimed to have recognised distinctly urban and non-urban lifestyles, yet these have been severely criticised (Glass 1989; Savage and Warde 1993: 2). However, classifications such as these are descriptive classifications. While they are able to describe what characteristics are present in urban areas, they are unable to isolate and identify anything that is unique to urban areas and which can therefore be used as a basis of a definition of 'what is urban'. These indicators identify differences of degree, not type.

For example, shops are found in both urban and non-urban areas. While the shops found in urban areas may be larger, more specialised, of a greater number and of a greater range than those found in non-urban areas, there is nothing exclusively urban about shops. The observable differences are differences of degree rather than fundamental differences of type. If a classification of urban and non-urban areas is devised on the basis of the shopping facilities found in each area, as indeed it has been, the decision of where to draw the boundary between urban and non-urban is not natural, or necessarily even obvious; it is a human decision. A classification such as this reflects nothing more than human opinion, not natural law. Indeed, it is the failing of many a classification that the two are often confused.

This fundamental uncertainty in classifications is reflected in the fact that classifications devised in different countries differ so much. In Scandinavia, for example, settlements as small as 300 people might be classified as urban, while in Japan a settlement must have 30,000 people to be classified as urban.

The question 'What is urban?', therefore, is resolved not by asking how many people, shops, offices make a city, but if the processes which form, transform, operate upon and operate within cities are unique to cities, or whether they are common to both urban and non-urban areas. The answer to this question is no. These processes are not unique to urban areas. The same processes form and transform non-urban areas as form and transform urban areas. While we may find a greater number or range of certain features in urban areas, they are not fundamentally different to non-urban areas in that they are the product of common social processes (Glass 1989: 56; Savage and Warde 1993: 2). Poverty, for example, considered a major characteristic of urban, particularly inner-urban, areas, is found to an equal, in many cases greater, degree in rural areas. Despite its presence in urban areas, poverty (like so many other things) is a social rather than an urban problem.

The apparently distinctive characteristics of urban areas are not, in fact, distinctive, only different. These differences merely reflect that social processes produce different outcomes in different areas under different local conditions. Given this, 'urban' should not be considered a natural, distinctive type of place, but merely a convenient label used by geographers and other social scientists to provide a spatial focus for their work.

Having said this, 'urban' is a meaningful label for a number of reasons. For example, geographers continue to organise research and publications around themes such as urban and rural. Therefore, libraries use 'urban' as a classification under which they file books and journals. In addition, categories such as urban and rural are reflected in the ways societies, organisations and governments organise and administer themselves. In the UK, examples have included the Council for the Protection of *Rural* England, *Urban* Development Corporations, The *Countryside* Agency, The *City* Challenge Programme, and so on. While 'urban' is not a natural, distinctive type of place, merely a convenient label, it is one that reflects social and academic perception and therefore forms a meaningful theme of inquiry.

What is urban geography?

Urban geography is, largely, what urban geographers do. While this may not be a very helpful point, it does reflect the subject's lack of precise definition. However, it is possible to recognise a number of concerns common to many urban geographers. These concerns may be summarised as being of three types. *Descriptive concerns* involve the recognition and description of the internal structure of urban areas and the processes operating within them or the relations between urban areas. *Interpretive concerns* involve the examination of the different ways in which people understand and react to these patterns and processes, and the bases these interpretations provide for human action. *Explanatory concerns* seek to elucidate the origins of these patterns and processes. This involves an examination both of general social processes and their different manifestations under particular local circumstances (Short 1984).

However, urban geography is nothing if not dynamic. The emphasis placed upon these concerns shifted significantly over the course of the twentieth century as urban geography developed as an academic discipline.

Urban geography has been characterised, over the course of its history, by a series of radical shifts in the way geographers have gone about investigating the urban. This is a reflection of radical shifts in the philosophies that underpinned these approaches. Each of these approaches has been characterised by a very different emphasis on the approaches outlined above. The following section reviews briefly the main approaches in urban geography in the twentieth century.

Changing approaches in urban geography

Early approaches

Site and situation

Studies from the early twentieth century were concerned primarily with the physical characteristics as the determining factor in the location and development of settlements. This concern has been long superseded in all but historical and some rural studies as cities have grown both in size and complexity. Original location factors have tended to be overridden by the scale of subsequent urbanisation or have greatly declined in importance as the form and function of urban areas have changed.

Urban morphology

This was an important root of urban geography. It developed particularly strongly in German universities in the early twentieth century. It was primarily a descriptive approach that sought to understand urban development through examination of the phases of growth of urban areas. Using evidence from buildings and the size of building plots, it aimed to classify urban areas according to their phases of growth. While this approach came in for some heavy criticism in the 1950s and 1960s as more scientific approaches came to dominate the subject and the social sciences generally, it made something of a limited come-back in the 1980s. Recent work has concentrated on the roles of architects, planners and other urban managers in the production of the form and design of urban areas (see e.g. Whitehand and Larkham 1992).

Modern approaches

The two approaches outlined above were associated primarily with the infancy of urban geography. A greater diversity and maturity was evident in the approaches that came to dominate in the post-1950 period. The most significant of these and their main critiques are outlined below.

Despite their obvious differences, the models outlined demonstrate some similarities. Basically they have all sought to examine the ways in which urban patterns and processes are the outcome of the combination of human choice and action and wider social processes which place constraints upon this human action. Consequently the approaches that follow have all aimed to explore three things. First, they have all considered the ways in which humans make choices about a variety of matters (e.g. where to shop, where to live, where to build, how to build) and the ways in which these decisions might influence urban patterns and processes. Second, they have all explored the constraints that might impinge upon this human choice and the ways this constraint might influence urbanisation. Finally, they have considered the outcomes of the relationship between choice and constraint. They have considered which is the dominant side of the relationship and the ways in which urban development is the outcome of the combination of choice and constraint. What distinguishes each of the following approaches is the relative importance they place on choice and constraint, and the ways in which they believe each to operate. Choice and constraint is a dominant theme of urban geography in the post-1950 period.

Positivist approaches

Although the positive philosophy dates back to the 1820s, it significantly influenced urban geography only from the 1950s. This reflected the impact of scientific approaches upon the social sciences generally and the increasing capacities of computers allowing the manipulation of ever more complex statistical datasets.

The positive philosophy is based upon the belief that human behaviour is determined by universal laws and displays fundamental regularities. The aim of positive approaches was to uncover these universal laws and the ways in which they produce observable geographical patterns. Positive approaches may be subdivided into two types – ecological approaches and neo-classical approaches.

Ecological approaches were based upon the belief that human behaviour is determined by ecological principles, namely that the most powerful groups, however this was defined (usually in terms of their income), would obtain the most advantageous position in a given space, for example, the best residential location. This school of urban geography dates back to the Chicago School of Sociology from the 1920s, and their contributions include Burgess' concentric zone model and Hoyt's sector model of land use (see Figures 2.1 and 2.2).

The ecological approach developed during the 1960s in that the models were refined with the increasing sophistication of computers. Despite this, they were able to offer little more than descriptive insights and by the 1970s were being criticised for their failure to say anything about the growing problems seen in cities; they started to be superseded by other approaches (Ley 1983).

An interesting application of these models was that developed by Mann in 1965 of the British city. The principal innovations of Mann's use of Burgess' and Hoyt's models was its combination of concentric and sectoral residential class patterns. This was based on the recognition that while cities might develop outwards in patterns approximating to concentric zones, frequently income differences are sectoral in pattern. Mann also incorporated the actions of the local authority in the provision of housing, an important aspect of the urban geography of post-war British cities, and the effects of negative industrial externalities on residential location. Mann argued that pollution from industry in the core was dispersed to the east of British cities by prevailing winds, thus degrading this area as a residential location. This created an east–west split in residential class. This was sustained by observation in a number of British cities. Despite these innovations, however, this model has dated poorly and is subject to all of the limitations that characterised this approach.

Neo-classical approaches were also based upon the belief that human behaviour is motivated primarily by one thing and is, therefore, predictable. However, neo-classicists believed that this driving force is rationality. By rationality they argued that each decision is taken with the aim of minimising the costs involved (usually in terms of time and money) and maximising the benefits (again time and money). This type of behaviour was referred to as utility maximisation.

The cities produced by positive models, of both types, were of neat, regular, homogeneous zones. Their very poor approximation to reality was the source of much of the criticism directed at these models, and reflected the overly simplistic assumptions upon which they were based and the important factors and motivations they ignored. Their failure to recognise and account for the idiosyncratic and subjective values that motivated much human behaviour was critiqued by behavioural and humanistic approaches that emerged in the 1970s and 1980s. These approaches placed the question of the complexity of human motivation at the centre of their inquiry. Positivist theories were also criticised for their failure to consider adequately the constraints within which human decision-making took place. A body of theories and approaches (given the umbrella title 'structuralist') which again emerged in the early 1970s sought to redress this imbalance.

Behavioural and humanistic approaches

Both of these approaches developed as criticisms of the failings of the positivist approaches. They were united in their belief that people, and the ways in which they make sense of their environment, should be central to their approach. However, they differed considerably in the ways in which they went about this. Behaviouralist approaches may be regarded as an extension of positivist approaches. They sought to expand positivism's narrow conception of human behaviour and to articulate more richly the values, goals and motivations underpinning human behaviour. However, despite this, they were still concerned with uncovering law-like generalisations in human behaviour. Behavioural approaches sought to examine the ways in which behaviour was influenced by subjective knowledge of the environment.

Humanistic approaches stemmed from a very different philosophical background. They sought to understand the deep, subjective and very complex relationships between individuals, groups, places and landscapes. In a radical departure from the scientific approaches of the 1950s and 1960s, the humanistic approaches brought techniques more associated with the humanities to understand people–environment relationships. This was reflected in the sources upon which they drew. These included paintings, photographs, films, poems, novels, diaries and biographies. The influence of humanism in urban geography was limited. Most humanistic work was conducted on rural or pre-industrial societies. Humanism developed in urban geography mainly as a critique of the monotonous, soulless landscapes of modern cities. The best developed application of the humanistic perspective in this regard is Edward Relph's *Place and Placelessness* (1976).

Both of these approaches, despite their differences, were less concerned with the production of descriptive models of urban form, and more with the production of interpretive insights into the relationship between people and their environments. However, limitations within the approaches themselves, and the criticisms from structuralists, again to do with their failure to consider the constraints acting upon human decision-making and behaviour, limited their long-term impact on the subject.

Structuralist approaches

Structuralist approaches in the social sciences generally, and in urban geography specifically, may be recognised through their conviction that social relations and spatial relations are either determined, or are in some way influenced, by the imperatives of capitalism as the dominant mode of production. This has led to criticisms that such analysis has failed to adequately account for the role of human action within these relations. Structuralist approaches have been accused of treating humans as mere passive dupes of economic structures. Much development in structuralist urban geography has involved the attempt to try to incorporate the 'structural' and the 'human' dimensions and thus overcome criticisms of 'reductionism'. Structural analysis in urban geography has largely derived from interpretations of the work of Karl Marx. Marx's view of history was of a series of 'modes of production', each of which was characterised by a particular structural relationship between the economic base and the social superstructure. Cruder versions of Marxism present the change in the economic base as controlling and determining change in the social superstructure. While this is a considerable over-simplification, the ways in which this relationship has been interpreted have varied a great deal over time and between different schools of Marxist scholarship.

The neo-Marxist influence on the social sciences in general dates back to the late 1960s. At this time there was a general call for geography to become more relevant, to help tackle and solve pressing social problems. This was prompted by militant reactions to issues such as the Vietnam War, urban poverty, racial inequality and increasing levels of debt among developing nations. It was felt that the quantitative, positivist geography had failed to address these problems. Quantitative geography was accused of naively ignoring the inherent consequences of the capitalist system, in particular the production of inequality. Neo-Marxist geography developed out of a critique of this system (Cloke *et al.* 1991).

Neo-Marxist urban scholarship has not formed a neat, coherent body of work despite deriving from a common ideological position (Short 1984: 3). Neo-Marxist urban geography has displayed both disputes between different authors and abrupt breaks within the works of single authors as they have whole-heartedly abandoned previous positions. The two most influential figures within neo-Marxist urban geography have been Manuel Castells and David Harvey, and their works demonstrate these tensions well (Bassett and Short 1989: 181).

Manuel Castells' two most influential books were *The Urban Question: A Marxist Approach* (translated into English in 1977) and *The City and the Grassroots* (1983). Both of these books were concerned with the relations between economic and social structures and spatial structures. *The Urban Question* provided a very abstract and theoretical reading of these relations, something for which Castells was frequently condemned by his critics (Bassett and Short 1989: 183). He was particularly concerned with the role of the state as a regulator of urban crises. These crises, he argued, following a well-worn Marxist tradition, derived from the contradictions inherent in the capitalist mode of production. *The City and the Grassroots*, as the name suggests, was a subsequent attempt to include human agency within his Marxist framework. This he attempted through a number of case studies of urban social protest movements and their influence within urban change. *The City and the Grassroots* was a recognition that dominant class ideologies and the imperatives of economic relations were not unproblematically stamped across space. Rather, spatial relations reflected the patterns of resistance and opposition that these imperatives met. For a more complete analysis of these relations between economics, society and space it was important that this resistance was recognised (see Bassett and Short 1989: 181–3).

The early work of David Harvey, for example *Social Justice and the City* (1973), represented an attempt to read historical cycles of urban development as a reflection of the resolution of crises of over-accumulation within various 'circuits of capital'. This is an approach that attempts to link urban restructuring to wider processes of economic restructuring. It focuses upon the built environment as a destination for investment, the profitability of which is linked to the state of the wider economy. Harvey argued that investment in, and hence production of, the built environment occurred when an over-accumulation of capital in manufacturing and commodity production caused returns in this sector to fall. This made land and property an attractive alternative investment.

Providing that a framework existed to facilitate it, these conditions caused capital to 'switch' from the former to the latter. Marxist terminology referred to capital switching from the 'primary' to the 'secondary' circuit. However, this capital switching and the patterns of urban growth tend to take on a cyclical form. Investment in the secondary circuit will tend to lead eventually to an over-accumulation in this sector, causing returns on investment to fall. The result of this is that capital will either switch back to the primary sector or seek more profitable investment opportunities within the secondary circuit. These are to be found in newer developments. This switching of capital is manifest in the built environment becoming abandoned in the wake of capital movement to more profitable opportunities elsewhere. Marxist economic theory applied by Harvey saw the built environment as the site for the temporary and somewhat unstable resolution of crisis of over-accumulation in the capitalist city (Savage and Warde 1993: 46–7). The examples that have been used to support Harvey's observations include the growth of post-war US suburbs and the office boom in the UK in the 1970s and 1980s.

However, this approach does have its limitations. Savage and Warde (1993: 48–50) in a review of Harvey's contribution to urban geography outline a number of these limitations. Harvey's account of both the built environment and social struggle is partial. It omits a number of crucial dimensions. Harvey suggests that capital switching within the secondary circuit involves a change of location. This is not necessarily the case. It ignores the conversion of property, for example factories into shopping arcades. However, just because capital may not actually switch location does not mean that these changes will not have social consequences. The workers made redundant from factories are unlikely to be those who get jobs in luxury shopping arcades and wine bars. Even if they did, the rewards would be unlikely to match those they received in their previous employment. Harvey's account of social struggle relies heavily on a conceptualisation of this struggle organised along the lines of class. While not necessarily theoretically flawed, this is at least limited. Harvey does not take much account of groups based on lines other than class. There are many cases of groups based on gender, ethnicity and sexuality having significant influence on the process of urban restructuring. For example, Manchester in the UK, and New York, Minneapolis, San Francisco and other cities in the USA have had gay groups who have been influential in the development of their inner areas (Lauria and Knopp 1985; Knopp 1987). Such alliances may cut across class lines.

Perhaps the most serious limitation of this approach is that it has failed to sustain any significant or wide-ranging research programmes. Harvey's own work has few detailed examples to back up his observations, and few others have stepped in to fill this vacuum (Savage and Warde 1993: 48). While Harvey's ideas have generated plenty of debate in academic circles, much of this debate has been very abstract and little has been rooted in concrete explorations of examples of urban restructuring. For example, theoretically some questions remain about Harvey's failure to distinguish the causes and consequences of capital switching. While these might normally be expected to be ironed out in subsequent applications of theoretical ideas, this has not been the case with Harvey's work (Bassett and Short 1989: 183–6; Savage and Warde 1993: 45–50).

Urban sociology

The relationship between urban geography and urban sociology has traditionally been close. The interchange of ideas dates back to the 1920s with the production of Burgess' model of concentric zones (Figure 2.1). This was the product of the work of the Chicago School of Urban Sociology that later formed a bedrock of research and teaching in urban geography. Urban sociology, like urban geography, has been far from static and has passed through a number of theoretical developments and debates. Urban sociology has been particularly influential in the practice of urban social geography. Some of the most influential work in urban sociology has stemmed from a body of work referred to as *neo-Weberian*, reflecting the influence of the sociologist Max Weber. This has offered a perspective on the city as a site of the regulation and allocation of scarce resources. Pioneering work in this vein was carried out by John Rex and Robert Moore, who investigated the concept of 'housing classes' through research into ethnic minorities and their access to housing in inner Birmingham. This resulted in the classic studies *Race, Community and Conflict: A Study of Sparkbrook* (Rex and Moore 1967) and *Colonial Immigrants in a British City* (Rex and Tomlinson 1979). In these studies they argued that people's access to housing was not simply dependent upon their job but upon a host of other factors, including ethnicity (Savage and Warde 1993: 68). It was argued, for example, that 'from the point of view of housing amenities, it is better to be a white labourer than an Asian chartered accountant' (Smith 1977: 233). Research uncovered examples of racism by private landlords, local authorities and building societies to substantiate these claims. Although in itself a diverse body of work that has evolved a great deal from its origins in the 1960s, it has highlighted the importance of the role of public and private 'gatekeepers' such as the local authority, estate agents and building societies, and the role of consumption rather than production in the creation and maintenance of social divisions (for a summary see Ley 1983: 280–323). The neo-Weberian approach has fallen out of favour as sociologists have found the ideas of social classes derived from both housing and employment situations to be too broad to be usefully used as an analytical tool because they conflate the essentially distinct areas of employment and housing (Saunders 1990; Savage and Warde 1993: 69).

The city unmasked: the symbolic meanings of the urban landscape

Since the mid-1980s the city has been subject to scrutiny from a diversity of scholars pursuing a series of closely related approaches that have been labelled by some as 'the new cultural geography'. Although by no means all geographers, some of the best work in this vein has been conducted, for example, by anthropologists and historians. Despite its diversity, scholars working within the new cultural geography have been concerned with unmasking the meanings of cities, landscapes or buildings. Cities are not just collections of material artefacts; rather, they are also sites through which ideologies are projected, cultural values are expressed and power is exercised. The meanings of cities, landscapes and buildings may be inscribed into them by their producers, architects, builders, planners and proprietors. These producers are situated within cultural and class contexts. Consequently these meanings are not simply idiosyncratic and individual, although the biographies of individual producers are important sources; rather, they are reflective of wider positions derived from class, capital, nation, religion or some other such cultural context. Alternatively, meaning can be projected upon cities, landscapes or buildings through their representation in a variety of media – paintings, novels, poems, advertising, photography, film and so on. Advocates of new cultural geography and cognate approaches have argued that we can think of the city and its components both as the visible manifestation of less visible or invisible processes and cultural positions, and as symbolic representations of these cultural positions. In defining the symbolic meanings of urban space, scholars have advocated approaches derived as much, if not more, from the humanities as from the social sciences (Cosgrove 1989). This is apparent in the metaphors that describe their endeavour, referring, as they do, to the city as 'text' and to their approach as 'reading' the landscape.

> A framework is required to show how and why cities develop particular meanings, and how these are constructed, interpreted and sustained. One approach is to think of the city as a text, in the same way that a novel or film might be a text. This text has certain authors, is constructed in a particular way by various procedures and techniques, has a series of meanings embedded within it and is subject to forms of reading.
>
> (Savage and Warde 1993: 122)

There have been various frameworks put forward to deconstruct or 'unpack' the meanings inscribed in or attached to urban spaces. Despite their differences these frameworks are all described as being broadly 'semiotic', a name given to approaches that have examined non-verbal systems of communication or signs.

Various commentators have argued that there are a number of levels of meaning reflected in the built environment. Amos Rappaport (1990), for example, recognised three. First is a high level of meaning reflecting world or cosmological views of producers, such as the design of cities being intended to reflect views religious about the structure of the universe. Second is a middle level of meanings reflecting the class or cultural positions of producers. This might include the height of a building reflecting the dominant position of a particular group in society, or their perceived dominance, or an attempt to suggest their dominance. Finally is a low level of meaning reflecting the everyday use of space. This level of meaning may not be captured in the design of spaces but in the ways in which people use spaces that might affirm or challenge the meanings inscribed in them. Mona Domosh (1989, 1992) outlined a related model of the meaning of city landscapes through her work on the skyscraper within the corporate landscape of late nineteenth-century New York, but one which argued that landscape interpretation should take account of the personal imprint of producers, and functional imperatives behind the design of urban space as well as the imprint of more distant economic, social and cultural changes. These, she asserts, are typically intimately linked.

> Many of the early skyscrapers in New York were built as much as statements of personal identity, business acumen and aesthetic legitimacy as they were to provide the space necessary for the incredible economic expansion of the city in the late nineteenth century.
>
> (Domosh 1992: 84)

There have been many sophisticated examples of work that has sought to examine the symbolic meanings of urban space. An influential early example was David Harvey's interpretation of the shifting and contrasting meanings attached to the Basilica of the Sacred Heart in Paris (1979, reprinted 1989) as the site was struggled over in the course of a troubled history. Other interesting works in this vein include James Duncan's (1990) reading of the landscape of Kandy in Sri Lanka as a reflection of royal power and religious cosmology, and Ed Soja's (1989, 1996) readings of the landscape of modern Los Angeles as a reflection of the military ideologies of the city's authorities exercised in the containment of its populations, the power of corporate capital to display itself and to exclude the disorder of the city, and the paranoia of the city's wealthy residents.

As Harvey's work has demonstrated, the symbolic meanings of urban spaces are not fixed but change over time as societies change around them. An obvious example is the meaning of statues of communist leaders in Eastern Europe. From being symbols of the absolute power of the Communist Party, these monuments became sites of contest during revolutionary uprising, and are now relic landscapes – landscapes symbolising a power that has passed, or who's original meaning has been lost. Social change may be rapid in the case of revolution, but gradual social change will also affect the symbolic meanings of landscapes. The increasing secularisation of British society and consequent gradual decline in church attendance has resulted in many town and city centre churches being closed down in the UK. Subsequent uses as pubs (Cheltenham) and nightclubs (Doncaster) have involved them taking on a very different set of associations and symbolic meanings than those originally intended, despite only limited alteration to their design and layout. With such shifts in mind Denis Cosgrove (1989) has proposed that we think of symbolic landscapes in terms of dominant landscapes, alternative landscapes, and also residual and relic landscapes.

Despite the obvious theoretical sophistication of much of the work that has sought to unmask the hidden ideologies and meanings inscribed into urban landscapes, it has recently been subject to criticism. Such work may be said to display a productionist bias. That is to say that in being concerned overwhelmingly with the production of urban meaning, namely its inscription into urban landscapes, it has neglected to say anything significant about how people understand, react to or 'consume' these meanings and ideologies. There is no necessary correspondence between the meanings inscribed into urban landscapes and those taken out as individuals encounter these landscapes. Indeed, the meanings that people take from urban landscapes are far more multiple and complex than much of the work in the semiotic vein would have us believe. The demolition of the myth that people are passive consumers of ideologies has made little impression on this work, with a few notable exceptions (e.g. Ley and Olds 1988; Duncan 1992; Ley and Mills 1993). Ley and Mills (1993: 258) criticise this productionist bias:

> Such models of hegemonic control present the consciousness of the masses as monolithic and unproblematic, passive and without the potential for resistance. The view of mass culture is distant and elitist. Soja's (1989) view of the surveillant state in Los Angeles, for example, is a view from on high; as the noose of total social control is drawn tightly around the city, we do not know if any member of the thousands of cultural worlds in that city has noticed, for no other voice or values are admitted other than those of the author. As in so much of the literature on cultural hegemony, the social control of consciousness is alleged but never proven. When we look for the voices of the manipulated masses we encounter in the text a gaping silence.

Such gaps have begun to be addressed by recent work on, for example, the landscapes and practices of shopping (Evans *et al.* 1996; Miller *et al.* 1998). Previous work on the shopping mall had often involved detailed readings of the architecture, ornamentation and layout of the mall as a reflection of the instrumental ideology of retail capitalism (see Gottdiener 1986). There was an implicit assumption that these landscapes were closed to any alternative interpretation or contestation. Analysis of the actions of shoppers in these environments was absent from such works. However, recent work has shown shopping to be a much more complex activity than previous work had suggested, the act involving multiple meanings to different groups of people and its landscapes being interpreted in a plethora of contrasting ways by shoppers. Theoretically this draws on ideas originating in literary theory about the process of reading. Ideas, such as Brian Stock's (1986) notion of the textual community, for example, suggest that people's interpretations of texts, be they literary texts or urban landscapes, are not entirely individual but rather that groups of people might come together around shared readings of texts. Such work has gone a long way in demonstrating the limitations of the semiotic mode of analysis and signals the beginning of a new direction in the analysis of the meanings of urban spaces.

Discussion topic

Consider the urban landscapes you encounter on a regular, everyday basis. Can you think of the meanings and uses that the producers of these landscapes intended? How are these meanings and uses written into the landscape, for example, through design, and possibly enforced, for example, through written rules and policing? Do you think the users of these landscapes know or understand these intended meanings and uses? Can you think of examples where alternative meanings are ascribed to these landscapes or where their use transgresses or resists the meanings intended by their producers?

Urban theory today

Urban theory in the 1990s was in a state of some uncertainty. No one philosophical perspective holds ascendancy. Indeed, one of the few positions that unites urban geographers is their wariness of the grand claims of totalising urban theories. One negative consequence of this has been the tendency among urban geographers to shy away from overt theoretical debate. However, on a more positive note, this lack of any single philosophical hegemony has opened up urban geography to the application of an eclectic range of perspectives. 'Readings' of the city are more likely to encompass perspectives derived from literary theory, film studies, psychoanalysis and cultural studies as they are Marxist economics or neo-Weberian sociology.

This eclecticism and cynicism of grand theory are a result of the feeling that the urban theories outlined in this chapter are unable to provide anything more than partial accounts of the city. Further, they are becoming increasingly remote from the new forces affecting the development of cities, for example new technologies, new forms of governance, new economic forces and new ecological concerns. It is unlikely, however, that totalising urban theories will ever vanish entirely. Indeed, it is to be hoped that they do not. Much of the intellectual development of urban geography has derived from the debates opened up by its clashing philosophical adherents. However, this is an appropriate point at which to argue that for urban theory to offer a useful contribution it must do two things. First, it must grow and change as cities develop. Theoretically informed accounts of the 'electronic' city and the 'sustainable' city are as imperative in the early twenty-first century as they were of the industrial city in the mid-twentieth century. Second, urban theory must be rooted around real urban issues. There are a range of continuing and very pressing urban problems affecting the world's metropolises. These will not go away as cities change. Contrasting perspectives on these rather than wildly abstract arguments can only enhance the subject's relevance.

This book does not attempt to develop any new or alternative totalising urban theory. Indeed, much of it is written wary of any such theory. Rather, it aims to offer a series of syntheses of significant bodies of literature on the emergent forces of urbanisation and the urban changes of the late twentieth and early twenty-first centuries.

Essay titles

- The development of urban geography has been characterised more by revolution than evolution. Discuss.
- To what extent and in what ways has the development of urban geography been influenced by contributions from other disciplines?

Further reading

Caves, R. (ed.) (2005) *The Encyclopedia of the City*, London: Routledge. [Contains concise and useful summaries and essays on key ideas and issues.]

Fyfe, N. and Kenny, J. (eds) (2006) *The Urban Geography Reader*, London: Routledge. [Companion volume to *The City Reader* (see below).]

LeGates, R.T. and Stout, F. (2003) *The City Reader*, London: Routledge (2nd edn). [Key reference work containing many original readings from key writers in urban studies.]

Pacione, M. (2001) *Urban Geography: A Global Perspective*, London: Routledge (ch. 2). [Provides a sound introduction and overview of the subject.]

Savage, M., Warde, A. and Ward, K. (2002) *Urban Sociology, Capitalism and Modernity*, London: Palgrave Macmillan. [An excellent guide to approaching cities, written predominantly from a sociology perspective.]

4 The changing economic geography of the city

Five key ideas

- **Study of the economic geographies of cities must recognise the context of the world economy.**
- **De-industrialisation has had a severe impact upon the economies of many cities.**
- **Newly emergent sectors of the economy have tended to display distinctive geographies to previous rounds of economic activity.**
- **Much economic activity is decentralising within and beyond cities.**
- **Cultural and creative industries have shown some potential to regenerate former industrial inner-city areas.**

Cities and the world economy

It has become clear that urban development is fundamentally influenced by position in the world economy. This raises important questions about how we understand this process. First, it suggests we cannot understand the processes that shape and reshape cities by looking only within cities. We must adopt a much wider perspective, one which recognises that cities are shaped by processes from far beyond their boundaries, as well as factors much closer to home. Despite this we must not lose sight of the fact that cities are not the helpless pawns of these processes. These global forces are mediated locally; namely their outcomes are determined by local factors such as the nature of local urban governments, economies and cultures. Cities are shaped by the interplay of local, regional, national and international forces (Healey and Ilbery 1990: 3–6).

Second, while the world economy is becoming increasingly interconnected through the international operations of multinational or transnational corporations, the international dimension of urbanisation is not new. Cities have long performed international functions and many have been profoundly shaped by these functions. Older industrial cities of the UK have long traded with countries all around the world. London was the command centre of a worldwide empire. The legacies of these international functions are still apparent in the landscapes, economies and institutions of cities as well as the links between cities.

Despite long histories of international functions there are a number of reasons why the recognition of the international contexts of urbanisation is of renewed significance. Previous accounts of urbanisation have tended to under-recognise the significance of the international dimension of urbanisation. They must be regarded, therefore, as partial accounts. This came at a time when the adverse effects of international competition were being felt in the economies of many of the older industrial cities of Europe and North America, international organisations such as the European Union were becoming increasingly significant shapers of fortunes locally, and an increasing number of influential companies were operating internationally (Knox and Agnew 1994; Hamnett 1995).

This chapter considers the international dimension of urbanisation through four questions:

- What have been the main trends in the world economy in the past thirty years?
- What have been the impacts of these trends upon the cities of Europe and North America? This will consider the impacts upon the internal structure and operation of cities and upon the relations between cities.
- How have these impacts varied between cities of different types?
- How have cities responded to these changes?

De-industrialisation and the city

Many of the major cities of North America and Europe were founded, or became closely associated, with the expansion of industrial capital from 1850 onwards. The rise of industry within cities represents a major phase in the history of the city. However, by the early 1980s the majority of these cities were experiencing severe problems with their economies. Unemployment, through the decline of the manufacturing sector, emerged as the major problem facing older industrial cities in North America and Europe. The rapidity and extent of this problem were startling (see Table 4.1).

Table 4.1 ***The decline of manufacturing employment in the UK 1975 to 2004***

Year	*Number of employees (thousands)*
1975	7351
1980	6801
1985	5254
1990	4994
1995	3918
2000	3951
2004	3282

Source: *Labour Market Trends/Employment Gazette*

Examination of the stark figures revealed that the problem of manufacturing decline in urban areas displayed a number of important dimensions:

- *Temporal* The emergence of a problem of long-term unemployment. Significant numbers of unemployed people remained out of work for periods of well over one year.
- *Sectoral* Unemployment was concentrated in manufacturing which was once a dominant sector of national economies.
- *Regional* Important inter-regional dimensions emerged. Areas such as the North of England and the manufacturing belt of the American Midwest emerged as regions with severe economic problems.
- *Urban* Cities, the focus of manufacturing industry, bore the brunt of manufacturing decline. This was largely concentrated in their inner areas.
- *Social* The worst impacts of unemployment were concentrated in a number of social groups including youths, the late middle-aged, males and ethnic minorities.

The de-industrialisation of the manufacturing cities of North America and Europe may be attributed to three factors: factory closure, the migration of jobs to other areas of the country or abroad, and the replacement of jobs by technology.

Impacts of de-industrialisation

Not all cities in North America and Europe were equally affected by the processes of de-industrialisation. Those cities with diverse economies or without a significant manufacturing component in their economies enjoyed very different economic fortunes during the period of de-industrialisation. It is also difficult to generalise about the impact upon different industrial cities. The impacts of de-industrialisation varied between different industries, and depended upon the particular composition of individual urban economies and national and local government actions. However, despite these qualifications it remains true to say that de-industrialisation was the most significant economic process to affect large cities in Europe and North America since the 1960s.

One of the most dramatic demonstrations of the loss of economic dynamism in the urban industrial sector has been the fall in the total urban populations of large British cities. This was due predominantly to out-migration, a process known as counter-urbanisation or the suburbanisation of populations beyond municipal boundaries. This was a particularly noticeable characteristic in the 1970s and early 1980s.

Cities did show some signs of recovering population in the late 1980s. This process displayed a pronounced social dimension with mainly middle-class residents moving to the suburbs or beyond, while more disadvantaged groups were far less mobile. Coupled with the migration of industry out of large urban areas and factory closure, this led to something of an economic vacuum developing in inner-city areas and a spatial polarisation of urban populations based around income, lifestyle and opportunities. Increasingly, the inner city found itself disconnected from the dynamics of the formal economy and developed, or failed, as a place and a people left behind, as these dynamics shifted elsewhere.

Explaining economic change

One of the most discussed causes of de-industrialisation has been the migration of jobs to newly industrialised countries. It has been argued that this represents the emergence of a new international division of labour in which the manufacturing functions of the inner areas of older industrial cities have been surpassed. This theory is able to account for a number of major, worldwide economic developments. These include the de-industrialisation of Western cities, the growth of cities in newly industrialised countries and the growth of global cities as the control and command centres of an interconnected world economy. However, despite this, the explanatory scope of this theory, while not being incorrect, is limited (Savage and Warde 1993). In relying so heavily on economic processes it is able to say little about, for example, the social geographies of cities that are clearly related to economic change. This theory is able to offer only one-directional explanations of the relationship between economic change and urbanisation. This perspective is able to demonstrate how macro-economic change has impacted upon cities. However, it can say little about how the characteristics of individual cities, their political leadership or local business community, for example, can affect the operation of macro-economic processes. It is able to look at the city from the outside in, but it fails to be able to see from the inside out. Cities do not find their position in the international division of labour automatically determined by their geographical location or their economic histories. They may be constrained by these factors but they are able to influence the operation of the economic processes affecting them. The new international division of labour perspective has little to say on these important aspects of the relationship between economic change and urbanisation (Savage and Warde 1993).

An alternative theory which adopts a far less abstract approach than the previous one is the 'restructuring' theory. The 'restructuring' refers to the restructuring of organisations in response to changing economic conditions. This approach was developed primarily by two geographers Doreen Massey and Richard Meegan in the late 1970s, and was refined during the early 1980s in publications such as *The Anatomy of Job Loss: The How, Why and Where of Employment Decline* (Massey and Meegan 1982) and *Spatial Divisions of Labour* (Massey 1984). Massey and Meegan noted that spatial unevenness in rates of employment and unemployment had come to characterise the UK economy. This, they argued, was the result of the spatial restructuring decisions undertaken by companies and organisations. Companies were able to use space (for example, differences in labour costs across space) to their advantage. Restructuring was organised around exploiting these spatial variations in an attempt to maintain profits in the face of an increasingly competitive world economy. Patterns of uneven national and international development reflected the ability of organisations to use spatial variations, not only in labour costs, but also in the socio-cultural characteristics of areas (for example, history of unionisation and militancy).

The de-industrialisation of the inner areas of older industrial cities may be interpreted as a restructuring response by manufacturing companies. These areas were characterised by high labour costs, and militancy compared to suburban and rural locations. Massey and Meegan described how regional specialisms developed as the production process was broken down into its component parts and became spatially dispersed. This was able to explain the emergence of the South East region of the UK as an area with a high concentration of headquarters and command and control functions, while older industrial regions became the sites of routine production in plants controlled and owned elsewhere. Massey and Meegan looked at the increasing mobility of capital and the increased salience of the social rather than the natural characteristics of areas.

This approach generated a number of positive advances. It sustained a major research programme into the changing urban and regional structure of the UK and was able to go beyond economistic explanations to show how social and cultural characteristics are implicated in economic restructuring. A major insight to stem from this was the demonstration that the relationship between economic restructuring and the specific geographical make-up of places was a two-way rather than a one-way relationship. Not only did economic restructuring impact upon the geographies of individual places, but also these unique geographies themselves impacted upon the operation of the processes of economic restructuring. The restructuring approach showed how places were not simply the passive receptors of economic change layered down from above but were active in affecting the outcomes of these changes (Savage and Warde 1993).

Despite this, the restructuring approach was the subject of increasing criticism during the late 1980s and 1990s. In contrast to other approaches the restructuring approach poorly articulated the impacts of economic restructuring upon specific places. Research in the restructuring vein has tended to suggest that places are far more coherent and heterogeneous than they actually are. The 'localities', as they were termed, in which this research was conducted were all relatively small and represented quite distinct areas. However, even these demonstrated a great deal of internal diversity, spatially, economically and socially. The impacts of economic restructuring upon these localities were therefore dispersed throughout the social and economic structure of the locality. The impact depended very much on one's location within this structure. It is not necessarily meaningful to adopt a spatial unit that suggests some degree of internal homogeneity to study these impacts. Overall, the restructuring approach tended to overemphasise the role of space in the mediation of economic restructuring and underemphasise the importance of the internal structure of the locality. The restructuring approach should be regarded as an ambitious but limited attempt to show the importance of social and cultural factors in the mediation of economic restructuring (Savage and Warde 1993).

Discussion topic

The relationship between cities and the world economy is a reciprocal one in which cities are active. With reference to the above discussion, think of ways in which cities are active in shaping their economic futures. Who are important agents within cities (individuals and organisations) in this regard? In what ways do they attempt to shape their economic futures and in what ways are they constrained?

Cities and the rise of the service economy

A great deal of optimism has been placed on the rise in service sector employment since the early 1980s. It was felt that the rise of this alternative sector might offset the losses experienced in the industrial sector in many cities of Europe, North America and Australia. However, the rise of services was sectorally, socially and geographically specific. Not only were they unable to fully compensate for the loss of manufacturing jobs but also the places and people who benefited from them were very different from those who bore the brunt of manufacturing de-industrialisation (Hudson and Williams 1986: 112; Allen 1988).

Service sector employment rose for a number of reasons in the early 1980s. These included the demands of businesses for specialised financial and legal services, the co-ordination required to orchestrate spatially dispersed economic activities within companies and the increased demands of households for services. Between 1945 and 1990 service sector employment grew by approximately 75 per cent. This growth was sectorally uneven, with particularly rapid growth in the distributive and banking and insurance producer services (Cameron 1980; Allen 1988; see Table 4.2).

The absolute rise in the number of service sector jobs has been significantly less than the absolute number of jobs lost in the manufacturing sector. This is reflected in rises in the total number of unemployed people. This is clearly demonstrated by the case of the UK economy, which, despite a significant rise in service sector employment, suffered a rise in the total unemployed between 1981 and 1991. What these absolute figures fail to demonstrate, however, are the resultant social and spatial dimensions of the change in the composition of employment.

The regional and urban impacts of the rise of the service sector and the broad shift from manufacturing to services within the UK economy, for example, may be interpreted as a complex interplay between social, economic, temporal and sexual dimensions. The broad impacts were an increase in the overall rate of unemployment, the transformation of local labour markets and the spatial decentralisation of the service sector across the UK.

First, the rise in service sector employment has been unable to compensate fully for the loss of manufacturing jobs. This broad shift displays a markedly uneven regional dimension. The worst affected areas in terms of overall increases in unemployment have been the urban areas of former manufacturing 'heartlands', for example in the Midlands (Spencer *et al.* 1986), the North of England and the Celtic 'fringe'. This has included areas such as South Wales, the West Midlands, the North East, Clydeside in Scotland, and Northern Ireland (Champion and Townsend 1990). Similar patterns have emerged in both North America and Australia. A distinctive rust-belt of north-eastern US cities emerged in which heavy manufacturing job loss has occurred. These cities include Baltimore, Pittsburgh, Cleveland, Detroit and Chicago. Economic growth in the USA has been largely spatially distinct from these areas of decline, focusing on an emergent sun-belt in, for example, California and involving industries using new technologies such as animation, biotechnology and space research (Knox and Agnew 1994: 247). Australia's rust-belt has included the states of Victoria, Tasmania and South Australia, while a sun-belt has emerged including Queensland and Western Australia (Stimson 1995). Sectoral economic shifts inevitably involve transformations in the 'space economy'. The use of the climatic metaphor sun-belt is appropriate to describe and explain the emergence of new sectors and areas in the space economy. Often the possession of a pleasant climate is one of the few natural advantages of significance in explaining the growth of new industries. The climate of, for example, California is one of the reasons why the highly qualified employees of new industries are attracted there.

Not only have rust-belt areas been quantitatively worst affected with the most severe retrenchment of male employment coinciding with sluggish service sector growth, but they have also been qualitatively transformed. The new service sector jobs have tended to differ significantly from the manufacturing jobs lost. The impacts of this on labour markets have included the rise of part-time working and flexible work practices such as temporary contracts and 'hire and fire' recruitment, and the increasing involvement of women in the labour force. Some implications of these changes have included the polarisation of income opportunities with the erosion of the intermediate income layer within labour markets and changing social relations within the home as women have progressively replaced men as the family breadwinner in many areas.

The geography of service sector growth is similarly complex. However, in the UK a major aspect of this may be summarised as the selective decentralisation of office employment away from the South East. The office sector in the UK developed rapidly in the 1960s and was heavily concentrated in London and the surrounding South East region. The office concentrations included the headquarters of national and multinational corporations seeking proximity to the financial district of the City of London, the headquarters of newly nationalised industries and the expanding Civil Service. The early decentralisation of offices was prompted largely by central government action. This included the Office Development Permit that encouraged movement beyond the South East, the Location of Offices Bureau (1964–79) which promoted alternative office locations and the movement of government offices away from London, for example the vehicle licensing office to Swansea (Hudson and Williams 1986: 111). This was a deliberate attempt to even out the development of office employment across the country. Private companies began to decentralise during the 1960s and 1970s. However, they tended to remain within the South West region, for example Eagle Star (insurance) which moved to Cheltenham and IBM (computers) which moved to Bristol. It was rare to find examples of significant relocations beyond this region.

Telecommunications improvements have created the possibility of the ‘virtual office’ (spatially distant but electronically interlinked networks of offices within companies) and companies have begun to spatially ‘disintegrate’ their offices according to function (Bleeker 1994; Graham and Marvin 1996: 128). This has led to some relocation of routine, ‘back-office’ functions and some middle-management functions to peripheral locations in former manufacturing cities and locations on the Celtic fringe. These relocations have taken advantage of cheaper and more flexible labour. Higher level management office functions have not relocated to anything like the same extent as more routine functions.

The types of jobs created by the increase in service employment have tended to be polarised between managerial jobs which are relatively small in number and more routine (back-office) jobs which are characterised by lower levels of pay, lower skill requirements, a lack of training and union representation, poor prospects and part-time or temporary contracts. This polarisation of opportunity has failed to replace the middle section of the labour market (well-paid, full-time, semi-skilled jobs) that was devastated by the process of de-industrialisation. Socially these opportunities have failed to re-incorporate those made redundant through de-industrialisation. The labour markets of Western economies have been characterised by a shift from male to female employment in former industrial areas as well as the national economies generally. The decentralisation that has characterised the rise of the service economy generally was fuelled initially by advances in telecommunications. However, the further advances in telecommunications threaten to undermine the advantages of peripheral areas in Western nations by allowing the further decentralisation of back-office functions abroad. Again the motivation behind this is the advantages associated with an oversupply of cheap labour in these areas (Graham and Marvin 1996: 153).

Case study A

Scottish call centre geographies

Call centres (or contact centres) – large centres that deal with the transaction of customer business by telephone – are an emergent element of the service sector in many regions. In the UK, for example, they were part of the widespread decentralisation of routine, back-office service functions to peripheral regions during the late 1990s. Since then this decentralisation has become international with an increasing movement of call centres to countries such as India and South Africa. Some media reports have suggested that the boom to regional economies from the development of call centres is to be short lived, since they are increasingly undercut on cost by overseas locations. Evidence from Scotland suggests that the reality is a little more complex than this.

There were 290 call centres in Scotland in 2002, one-third of which are in financial services. These employ some 56,000 people, 2.3 per cent of Scotland's total workforce. Most of these centres are located in Glasgow (and its hinterland) and Edinburgh, with smaller concentrations in cities such as Greenock and Dundee, and 40 per cent of them are owned by Scottish companies. The decisions made by companies to open Scottish call centres were partly based on availability of appropriate labour, financial assistance by government agencies and the Scottish origins of many companies.

While some companies have begun to relocate certain call centre operations overseas, principally to India, this does not necessarily represent the death-knell of the Scottish call centre industry. To date, the types of jobs being relocated overseas have tended to be routine, repetitive functions, though more complex jobs, for example, technical support, are beginning to follow. The major destinations for these relocated jobs have been the large cities of Delhi, Mumbai and Bangalore. Call centre jobs are being relocated to these destinations for a number of reasons, including reducing costs, availability of skilled labour, greater operational flexibility and extended customer contact hours. Companies have, however, citied problems with relocations, including language and cultural differences, difficulties with managing distant operations and poor customer satisfaction, along with high start-up costs.

Case study A continued

Despite the trend towards relocation, companies do not envisage a straight Scotland–India locational trade-off. Rather, companies make decisions within wider strategic frameworks. Factors influencing decisions include the efficiency and performance of overseas centres, the state of national economies and the world economy, and the ties between a company and its home base. The interrelations between these factors and their outcomes in terms of job losses and relocations are difficult to predict.

While it is likely that call centre jobs in Scotland will continue to be lost to India and other overseas locations, the number, and the scale of threat to the industry, is hard to assess. What is likely, however, is that the types of jobs likely to leave Scotland to go overseas will be routine, repetitive functions in large financial services companies, particularly where mergers and rationalisation have occurred.

Source: Taylor and Bain (2003)

Emergent economic geographies of the city

This section examines the ways in which the urban economy is becoming reconstituted around new or growing sectors. It considers the implications of this for different types of cities and for the internal geographies of cities.

Cities and corporate headquarters

Corporate headquarters have always displayed an urban bias in their location. While this has remained, except for some national corporations and some small multinational corporations which have relocated to suburban locations or smaller urban settlements, locational change has occurred reflecting shifts in the economic geography of national economies. Manufacturing cities and regions have tended to decline as centres for corporate headquarters. This has been most noticeable in North America where the cities of the Midwest have lost headquarters functions. There has been a general shift in the location of corporate headquarters in Europe and North America from manufacturing cities to those more associated with service economies (Knox and Agnew 1994: 251).

The headquarters of large multinational corporations have tended to concentrate in the very largest cities, and have become disproportionately concentrated in a small number of 'world cities' or 'global cities'. London, Tokyo and New York are the most advanced of this numerically

Table 4.2 ***Global cities and corporate headquarters***

City	*Fortune Global Service 500[a] Headquarters*	*Fortune Global 500[b] Headquarters*
Amsterdam	4	1
Atlanta	5	3
Chicago	7	10
Frankfurt	8	3
London	28	35
Los Angeles	7	10
New York	25	12
Osaka	20	21
Paris	28	23
Sydney	4	5
Tokyo	86	83
Washington, DC	6	5

Source: Johnston *et al.* (1995: 238)

Notes: [a] includes banking, financial, savings, insurance, retailing, transport and utilities services
[b] includes all industrial companies

small but hugely influential group. Table 4.2 indicates the concentration of corporate headquarters in these cities (Sassen 1991, 1994; Knox 1995).

The consolidation of corporate headquarters in large urban areas reflects their need for access to regional, national and international markets, a highly skilled labour force and a range of sophisticated, specialist service inputs. Only the very largest and well-connected cities are viable to satisfy these requirements. Corporate headquarters in these cities have become the focus of new, dynamic economic quarters that have begun to shape both the central areas and the wider economies of these cities based upon the growth of business (producer) services (Sassen 1991, 1994).

Producer service economies

Producer services offer legal, financial, advertising, consultancy and accountancy services to companies. They have become increasingly important to companies which need to respond rapidly to changing market conditions which require that they draw upon a range of specialised inputs. These services have become the fastest growing sectors of national economies and the economies of large cities in Europe and North America with significant numbers of international linkages (Thrift 1987; Sassen 1991, 1994; Hamnett 1995; Knox 1995).

Producer services have been overwhelmingly concentrated in the central areas of major world financial centres. They draw on the highly innovative environments of these cities and the extensive and instantaneous links to other industries and experts they offer. As well as the physical and electronic infrastructures of these sites, the social environments of bars, restaurants and health clubs are particularly important to the formation of social networks which are heavily drawn upon in business. This social milieu is impossible to reproduce outside these tight agglomerations of related activities even with sophisticated telecommunications links (Sassen 1994; Graham and Marvin 1996).

The main customers for producer services are the headquarters of multinational corporations. As the increasing extent and complexity of these corporations have proceeded, an effective central co-ordination, control and command complex has become ever more crucial. This effectiveness has come to depend to a large extent upon the use of specialised producer services. Producer services have become a vital part of the co-ordination of an internationally dispersed but increasingly interlinked global economy. The concentration of corporate headquarters in global cities has provided them with a lucrative, spatially circumscribed market.

Producer services have considerable impacts on the wider urban economies of which they are part. Producer services, along with certain financial and investment services, enjoy a privileged position in their urban economies due to their ability to generate 'superprofits'. Superprofits occur when disproportionately high returns are generated relative to investments of time and money. Because of this, producer services are able to dominate competition for land, resources and investment at the centres of large cities. As other sectors are squeezed out of both the spaces and the lucrative economies of these city centres they become devalued and marginalised. This has a significant social impact because the lucrative opportunities afforded by the producer service sector tends to be limited to a few highly qualified professionals. The remaining jobs in this sector are low-paid, low-skill jobs such as cleaning and security services. This polarisation of opportunity contributes nothing to the alleviation of the disadvantaged position of many inner-city residents. Indeed, the impacts may even be negative as locally oriented services are displaced by those such as boutiques and restaurants aimed at producer service workers (Sassen 1994: 54).

Research and development units

The location of company research and development units is dominated by two requirements: access to highly qualified personnel, and proximity to either corporate headquarters or production units. The former tends to exert the greater pull as research and development has become more central to the operation and development of companies in an unstable economic environment (Malecki 1991; Knox and Agnew 1994: 251–2). These requirements tend to be satisfied in one of three locations: large cities with corporate headquarters, near universities or other research institutions – particularly where a pleasant environment acts as a further attraction to qualified staff – or manufacturing sites. The last site reflects the need, present in some industries, to integrate closely research and development and production. As production has decentralised there is some evidence of associated research and development decentralisation (Knox and Agnew 1994: 252). Should this continue on a significant scale it is likely to rewrite the geography of 'branch-plant' industrialisation, expanding the range and type of linkages between branch-plants and their local economies (*The Economist* 1995). However, it is likely that this decentralisation will be highly selective. Branch-plants in peripheral countries are likely to remain simply as production units, disadvantaged by a lack of indigenous educational and research infrastructure.

The geography of research and development plants is highly significant in urban and regional development. Research and development complexes are sites of innovation where products are modified and new ones developed. They are potentially seed-beds for new patterns of economic growth (Healey and Ilbery 1990: 111–12), a fact recognised by local authorities who eagerly try to encourage the formation of innovative milieux through developments such as science and technology parks, linking universities and business. The problems faced by most of these suggest it is unlikely that research and development facilities are likely to disperse to any great extent except in association with branch-plants. In the future it is probable that existing centres are likely to consolidate the advantages they already possess.

New industrial spaces

A number of economic activities are beginning to emerge that are mediated through and based upon new technologies including e-technology and biotechnology. The nature, organisation and markets of these activities are sufficiently different from those of previous eras for them to be considered new industries. With new industries come new geographies of industrial location sufficiently different from previous forms of industrial locations for us to talk of 'new industrial spaces' (Knox and Agnew 1994: 252–3; Graham and Marvin 1996: 158–9).

These new industries are dominated by three locational requirements: the need for access to a highly qualified and functionally flexible labour force, an environment that facilitates constant innovation and cross-company and industry co-operation and good infrastructural and telecommunication linkages to corporate headquarters, universities and other research institutions, linked companies and national and international markets (Graham and Marvin 1996: 158–9). These industries tend to be attracted by the social, rather than the natural, characteristics of places.

Far from offering hope for the re-industrialisation of problem areas in older industrial cities, the locational and economic characteristics of these industries suggest they are likely to be harbingers of problems as well as possibilities. New industrial spaces are opening up in very different locations to old industrial spaces. The economic regeneration stemming from these developments is unlikely to directly or indirectly benefit those areas which have suffered from de-industrialisation. Emerging trends in industrial and economic geography suggest there is little chance that the dynamic economic mechanisms similar to those associated with industrialisation will return to inner-city locations.

Like other emergent sectors of urban economies, new industries appear to be dominated by a polarised income distribution requiring a small number of highly paid specialists and a large number of low-paid workers. Opportunities extended by this wave of development to those disadvantaged by the abdication of the industrial sector from urban areas are likely to be restricted to the latter. Whereas in the past the 'career-ladder' mechanism offered a pathway for career advance, there appears to be no evidence that low-paid and high-paid sectors in these industries are linked (Markusen 1983; Knox and Agnew 1994: 253). Working practices in these industries are likely to be dominated by the requirement to be flexible. They are also likely to be dominated by part-time working and short-term or temporary contracts. The effects of these new industries on the socio-economic welfare of disadvantaged groups is likely to be negligible, even negative.

Finally, there are question marks over the extent to which these industries are able to generate positive economic linkages to their local areas. These industries generate more linkages within companies and with national and international economies than they do with their local and regional economies (Turok 1993).

Cultural industries

One of the key responses of cities and central governments has been to try and attract new industries into declining inner-city areas. One sector that has been particularly vigorous in the regeneration of such areas has been the cultural or creative industries (O'Connor and Wynne 1996). These are industries such as fashion, media, film and video production, design, creative technology-based activities and music. These industries have the potential not only both to economically and culturally regenerate run-down inner-city areas but also to combine this with community-oriented regeneration projects based around, for example, new media, the internet or the creative arts such as music. Cultural quarters or spaces dedicated to the development of these industries have become a common feature of cities throughout Europe and the USA.

It is difficult to generalise about the development of the cultural industries sector in cities, as it displays enormous diversity. For example, some spaces are the result of organic development as artists or small new media companies moved into abandoned industrial districts to take advantage of extensive empty properties and low rents. This may be seen with the development of sectors containing high numbers of internet companies in former industrial districts of New York and San Francisco. In other cases the development of cultural quarters is a carefully managed aspect of urban regeneration programmes originated and run by local, regional or central governments.

Case study B

From custard factory to culture factory

An innovative example of the culture-led regeneration of the run-down inner city of Birmingham, UK may be found in the nineteenth-century Birds Custard Factory located in Digbeth, just outside the city centre. Digbeth was blighted by a lack of vigorous economic activity and a poor environment of warehouses and transport-based industry. The Birds Custard Factory had employed 12,000 people until it closed down in 1967. The site was empty until 1995 when an urban regeneration consortium was asked by the City Council to encourage growth of media and cultural industries in the city.

The consortium renovated the Birds Custard Factory and turned it into a complex consisting of over 150 studios, bars and restaurants and a small industrial

Case study B continued

development. Careful monitoring of small businesses moving into the site has ensured that an artistic and creative ambience has been maintained. Businesses located in the site now include: artists, fashion designers, photographers, models, furniture designers, recording studios, musicians, publishers and broadcasters. The aim has been to encourage businesses at the start of their careers and those who will benefit from strong synergies with other creative industries (Montgomery 1999: 32–5).

The site in itself has been a success. The renovation was successful, capitalising on the excellent industrial architecture typical of nineteenth-century factories. The studios were quickly filled and remain in high demand, and the Custard Factory is now the largest single creative complex in Europe (Fleming 1999: 15). The development has brought a much-needed buzz to what was previously a very drab inner-city area. Supporters claim that the development has started to create spin-off benefits in the surrounding area, encouraging students and affluent city dwellers to use the bars and clubs. The development also includes programmes aimed at combating social exclusion, for example, the *Artists in the Community* programme run by the University of Central England.

(Fleming 1999: 16)

Cities and telecommunications

The world economy appears to be becoming increasingly interdependent. The fortunes of individual places are not autonomous; they are increasingly bound up with the fortunes of other places and with processes operating at wider geographical scales. This increasing interdependence within the world economy is the result of a number of related processes: the emergence of multinational corporations as major shapers of international economic flows, the concentration of international command centres in global cities, and the deregulation of national financial markets and telecommunications advances 'gluing' together spaces and creating the 'space of flows' between cities in the world economy.

The most noticeable effect of this is that certain spaces are moving closer together, and interactions between them are becoming ever more instantaneous and 'real' (Hamnett 1995; Knox 1995; Leyshon 1995). Advances in telecommunications – telephone, fax, e-mail, computer networks and the internet – have reduced or eliminated the time delay in communication between distant spaces and increased the sophistication of exchange available. However, this 'electronic economy' is predominantly located in a small number of major cities which contain

Table 4.3 ***Urban dominance of telecommunication investment and use***

City	*Percentage of national population 1994*	*Measure of percentage of international communications infrastructure 1994*
New York	6	35[a]
London	16	30[b]
Paris	18	80[c]
Tokyo	10	37[d]

Source: *Financial Times* (1994), cited in Graham and Marvin (1996: 133)

Notes: [a] percentage of US outgoing calls starting in New York
[b] percentage of UK's mobile calls made within London
[c] percentage of French telecoms spending
[d] percentage of Japanese telexes in Tokyo

high concentrations of these technologies and which are highly interconnected with others. Again it is the global cities such as London, New York, Tokyo and Paris which dominate. As one moves down the urban hierarchy, concentrations of advanced communications technologies decline, as does the degree of interconnection with the world economy. These patterns show a high degree of primacy. As Table 4.3 shows, the largest cities within individual nations tend to account for a disproportionate percentage of that nation's international communications infrastructure (Sassen 1991, 1994; Knox 1995).

While the high concentrations of international, electronic communications infrastructure mean that a few select places are moving closer together, its converse is that a vast number of spaces with low concentrations of this infrastructure are failing to do so. This exclusion distances a large number of older industrial cities, cities in newly developing and less developed countries and rural areas from the centres of the global economy, which is effectively moving them further apart (Sassen 1994; Leyshon 1995). This is creating a geographical pattern of a small number of 'information-rich' global cities forming a highly interconnected transnational urban system, surrounded by vast 'information-poor' hinterlands with which they are poorly connected. It has been argued that as the interconnections between cities within transnational urban networks increase, the connections between these cities and both their regional hinterlands and domestic national urban systems decrease (Sassen 1994). While this is clearly an important emerging pattern, it would be premature to say that these cities are *more* connected to each other than they are to their own regions and nations. Traditional internal linkages between these cities and their regions are being complemented, and in some cases replaced, by flows of money, information, power and people from international cities (Hamnett 1995; Knox 1995). Whether they become usurped by them remains to be seen.

The geography of the international, electronic communications economy displays spatial, urban, sectoral, social and sexual divisions. This economy is overwhelmingly associated with the nations of North America, Europe and Japan. The extent to which cities in newly developing and less developed countries are linked is very limited. Linkages may exist in their capital cities but these are usually of a very limited extent, and are generally of a subordinate nature to the international economy, rarely being significant shapers of the geography of international flows. Within North America, Europe and Japan this economy is located primarily in cities of the service economy. Older industrial cities find themselves poorly connected and are either only the receivers of flows rather than the origins and the shapers of them, or are peripheral to them entirely. Access to advanced telecommunications tends to be highly restricted socially. They are associated primarily with a small number of professional, managerial jobs. Consequently a large proportion of the population form a massive information-poor underclass. This social dimension has a spatial expression with the inner city being the most obviously information-poor environment, an 'electronic ghetto'. Finally, the majority of jobs associated with the centre of the global information economy are occupied by men. This is, consequently, a very male space.

Discussion topic

What is the relationship between the geographies of past and present waves of economic activity? Do you think recent waves of economic activity have created new economic geographies of the city? Can you think of examples to support your view?

The 'urban doughnut'

Geographically the pattern of spaces of economic dynamism and economic depression appears to have become increasingly polarised. City centres have suffered a very mixed fate. The centres of a small number of global cities have boomed, fed by the rapid growth and superprofits of the financial and producer service sectors. The centres of some former manufacturing cities have been physically transformed through massive investment in convention centres, offices, hotels and retail and leisure developments (Sassen 1994: 43). Whether this has provided a sustainable basis for the economic regeneration of cities has looked increasingly precarious in the face of heightened inter-urban competition and problems such as periodic slumps in commercial property markets. Other centres, having failed to harness any mechanisms of economic dynamism, have declined along with the status of their cities. Historic cities have relied on their traditionally strong tourist and visitor economies to stay vibrant (Page 1995). However, the story of the inner city has been one of almost total and general decline, excepting those areas that have been subject to gentrification. 'Inner-city' conditions and problems have also become reproduced on a number of peripheral municipal housing estates in cities such as Liverpool and Glasgow. The future growth areas appear to be suburban or ex-urban, new metropolitan spaces beyond the boundaries of existing cities, discernible, for example, in the counties surrounding Los Angeles in California. The future of the economic geography of the city appears to be one of increasing decentralisation mediated through transport and telecommunications advance and change surrounding an inner city becoming progressively disengaged from the formal economy. This is what is meant by the term 'urban doughnut'.

Conclusions

This chapter has highlighted, at the broadest scale, the relationship between macro-economic change and urbanisation. It has also outlined some of the outcomes of the new micro-economic geographies of the city. Of major importance to urban geography are the ways in which cities have responded and tried to combat the negative impacts of economic change. These responses have been diverse and their implications wide-ranging. Attention turns in the chapters that follow to these changes and their impacts on the landscapes, economies, images and social geographies of the city.

Project idea

What evidence can you find of new economic sectors in a town or city with which you are familiar? In what areas of the city are they located? In what ways does the economic geography of your town or city vary from that of previous rounds of economic activity? Collect data and information and review local economic development policies to answer these questions. What are the implications of these new economic geographies of the city for those populations involved in previous rounds of economic activity?

Essay titles

- To what extent are many cities still trying to deal with the problematic legacies of de-industrialisation?
- '[Many believe that] the next industrial revolution will be based on the marriage of artistic imagination and technological innovation' (Hall 1995). In what ways will these forces shape the economic geographies of cities in the future?

Further reading

Bridge, G. and Watson, S. (eds) (2002) The Blackwell City Reader, Oxford: Blackwell (Part II, 'Reading urban geographies'). [A good collection of readings, many from classic sources.]

Dicken, P. (2003) *Global Shift: Reshaping the Global Economic Map in the 21st Century*, London: Sage (4th edn). [A classic exploration of the global economy.]

Marcuse, P. and Van Kempen, R. (eds) (1999) *Global Cities: An International Comparative Perspective*, Oxford: Blackwell. [Examines the social geographies of global cities.]

Paddison, R. (ed.) (2001) *The Handbook of Urban Studies*, London: Sage (Part IV, 'The City as Economy'). [An excellent set of essays on various aspects of the urban economy.]

Sassen, S. (2000) *Cities in a World Economy*, Thousand Oaks, CA: Pine Forge Press. [An account of the impacts of the global economy on the social geographies of cities.]

Web resource

National League of Cities – www.nlc.org

5 Urban policy and regeneration

Five key ideas

- **Urban regeneration is intended to ameliorate against the negative consequences of urban decline.**
- **Interpretations of urban problems and their causes shape urban regeneration programmes.**
- **It is important to situate urban regeneration programmes within a number of historical, geographical and political contexts.**
- **The nature of urban regeneration has changed markedly through time.**
- **The regeneration of urban areas has been 'partial'.**

Introduction

The idea of regulating capitalist cities emerged in the late nineteenth century. The disastrous consequences of unregulated urban growth became apparent to politicians and reformers in terms of health and quality of life in large cities throughout Europe and North America. It was within this historical and geographical context that formal planning systems began to emerge. If the planning system is thought of as a regulatory framework guiding the development of places, we can also recognise a more proactive set of interventions designed primarily to ameliorate against the negative consequences of urban decline. These interventions, termed 'urban regeneration', seek either to support vulnerable communities and localities through the redistribution of resources to them on the basis of need, or to promote growth and development and through this improve the lot of those most in need. This chapter aims to provide both an overview of major trends in urban policy and regeneration since the 1970s but also to outline a framework through which urban policies, regeneration initiatives and individual projects may be studied.

It is worth pausing to consider exactly what the aims of urban regeneration have been. Despite the huge variety of urban regeneration initiatives and policies enacted around the world it is probably true to say that they have all aimed to achieve one or more of the following four goals:

- Improvements to the physical environment (which have more recently come to focus on the promotion of environmental sustainability).
- Improvements to the quality of life of certain populations (for example, through physical improvements to their living conditions or by improving local cultural activities or facilities).
- Improvements to the social welfare of certain populations (by improving the provision of basic welfare services).
- Enhancement of the economic prospects of certain populations (either through job creation or through education or reskilling programmes).

Often regeneration has pursued more than one of these goals within any one policy, programme or project. In many cases policies have been pushed forward on the sometimes questionable basis that there exist causal links between some of these goals. For example, at times policies, whose explicit aim is economic development, have been advocated on the basis that they will also lead to improvements to the quality of life within localities undergoing regeneration. Finally, the relative emphasis on the goals of regeneration has shifted over time. For example, it is only since the early 1990s that environmental sustainability has risen substantially up many urban regeneration agendas. At the very broadest level we may argue that the aims of regeneration have broadened over time, from a concern primarily with living conditions to include economic, social and environmental goals.

A framework for analysing urban policy and regeneration

The questions below provide a framework to guide the analysis of either specific projects of urban regeneration or the more general policies or programmes of which they are part. Depending on the focus of any study and upon the nature of the case being examined, not all questions will necessarily be relevant in every case. However, they highlight key themes and issues to consider in the interrogation of urban policy and regeneration, and are used to frame the discussion that follows in this chapter.

A framework for analysing urban regeneration: twelve key questions

1 Urban problems

- What urban problem, or problems, have been identified?
- What has been identified as the cause of the problem or problems?

2 Policy contexts

- What is the origin of the policy/programme/project?
- What is its relationship to earlier approaches or those being implemented elsewhere?

3 Funding

- Where does the funding for the policy/programme/project come from?
- In what way is funding allocated?

4 The nature of regeneration

- In what ways does the policy/programme/project seek to achieve its aims?
- What are the outcomes of the policy/programme/project?

5 Stakeholders

- Who are the stakeholders involved?
- What are the relationships between the stakeholders?

6 Impacts of regeneration

- What are the impacts of the policy/programme/project?
- In what ways has the policy/programme/project been evaluated?

Urban problems

Urban regeneration policies are designed to address urban problems. However, urban problems do not present themselves to policy-makers; rather they are defined in various ways. The definition of urban problems has an important role to play in the legitimisation of urban regeneration and policy. While acknowledging that urban problems are real, as well as rhetorical, their articulation may be thought of as playing crucial roles within urban regeneration discourses.

It is generally recognised that urban problems are multifaceted. Urban regeneration has long recognised this, but has tended to define a particular problem, or a dimension of the many problems affecting urban areas, as the root cause of their woes. Commentators have recognised that urban problems include various combinations of environmental problems (such as derelict land, redundant industrial capital, inadequate housing stock, pollution and contaminated land), social/cultural problems (a lack of social cohesion within communities, crime, antisocial behaviour, poor schools and other public facilities) and economic problems (long-term, structural unemployment, a lack of indigenous economic dynamism). While multifaceted urban problems have long been characteristic of many urban areas, the focus of urban policy has shifted through time.

Inherent in all urban regeneration programmes is some articulation of the causes of the problems they are seeking to address. It is this articulation of causation that shapes regeneration. Put simply, urban regeneration is designed to address whatever policy-makers or practitioners think, or want to believe, is causing the problems they observe. As any student of urban geography will tell you, however, the causes of these problems are the subject of some debate. While organised urban interventions to improve the living conditions in cities can be traced back to the Victorian era, what we might now recognise as urban regeneration emerged in the 1960s. During this period it sought to address multiple deprivation through the spatial targeting of resources on a demonstration of need basis. This was predicated on the basis that prevailing wisdom located the problems of urban areas in failings of the people who lived in them, often referred to as a social pathology approach (Cochrane 2000: 535).

Subsequent interpretations problematised the notion that you could draw a line around 'problem areas' and explain their problems by simply looking within them. In the 1970s more structural explanations of inner-city problems emerged. These explanations posited that the problems of such areas stemmed from the consequences of economic structural adjustment. These explanations gained ground within academic circles during the 1970s, a reflection of the ascendancy of structuralism and Marxism within social science disciplines. At broadly the same time a reinterpretation of the structural thesis emerged within policy circles. Shorn of any associations with Marxism that might suggest labour is inherently disadvantaged within a set of capitalist relations, politicians began to acknowledge that the restructuring of industrial capital was at the root of many urban problems. The 1977 UK White Paper on inner cities and the subsequent Conservative government of Margaret Thatcher, elected in 1979, argued that urban problems stemmed from a lack of economic dynamism within localities. A raft of policies was designed throughout the 1980s to restore the fortunes of declining urban areas by luring the private sector back into them. Urban development corporations and enterprise zones were two of the most high-profile UK policies. This paralleled a shift towards what was called 'privatism' within American urban policy.

The 1990s saw a continuation of a structural explanation for the persistence of urban problems, this time couched more in terms of a failure of localities to compete within a globalised, post-industrial world economy (Oatley 1998). Consequently the early 1990s saw a number of policies designed to make localities, and businesses therein, competitive within the global arena. Initiatives such as City Pride (launched in 1995 in the UK) and based around the institutionalisation of place marketing was typical of the focus of urban policy during this era.

Since 1997, and the election of Tony Blair's 'New' Labour government in the UK, urban problems have been predominantly characterised in terms of social exclusion, the lack of social and economic relations between certain localities (for example, inner-city areas and peripheral housing estates) and the rest of society. Urban policies since 1997 have aimed to 'Bring Britain Together' (the title of the National Strategy for Neighbourhood Renewal published in 1998 by the Social Exclusion Unit). However, while the shift towards community regeneration is to be broadly welcomed as recognition of the failings of some earlier, more economically driven, policies, these more recent policies have drawn criticism on the basis of their definitions of the causes of urban problems. The Blair government has tended to see urban problems as stemming from 'operational' failures, for example, poorly run schools or the failure of other local institutions. What this definition side-steps are more radical interpretations that might put urban problems as stemming from unequal distributions of resources and opportunities (Imrie and Raco 2003: 30). As a result, the Blair government has sought to alleviate urban problems predominantly through policies that address operational failures rather than seek to address imbalances in the distribution of wealth. The latter course seems to represent an approach deemed too radical for contemporary mainstream political parties and policy-makers to entertain. Diamond argues: 'debate about wealth distribution has been shifted to the margins of regeneration by a fixation on operational issues' (2001: 277, in Imrie and Raco 2003: 30). Some critics have argued that this failure of policy-makers to engage with the question of wealth distribution as a cause of urban problems represents a 'weak' conception of social exclusion (Byrne 2001).

Policy contexts

Urban regeneration does not emerge within a vacuum; rather it is reflective of a number of contexts. Most obviously it is influenced by initiatives active in earlier periods or in other places. One aspect of the analysis of urban regeneration is to situate individual policies or initiatives historically (as part of an ongoing process of policy development through time), geographically (by tracing the influence of policies and practices from elsewhere) and ideologically (as reflective of prevailing political ideologies). The influence of these contexts is apparent in the case of urban regeneration in the UK since the 1970s. While it is tempting to characterise urban regeneration in a series of distinct periods, this tends to underemphasise the degree of continuity between periods. However, as a pedagogic devise this periodisation of urban regeneration does have some merit (see Table 5.1).

In the 1970s urban regeneration in the UK was concerned primarily with the physical renewal of older urban areas, typically within the 'inner city' that had emerged as a problem area in this period. However, a number of other issues came together in the late 1970s that led to urban regeneration responding to a wider range of imperatives. The late 1970s was a period of mass de-industrialisation in many older urban areas (see Chapter 4) that devastated the economies of many cities within industrial heartlands in Europe and the USA. The term the 'rust-belt' entered common parlance to describe regions undergoing this process. Cities in these areas were rapidly forced to confront the issue of having to find alternative sources of employment for large sections of their populations or to face the consequences of long-term, structural unemployment. At the same time a new political ideology was beginning to take hold within central government in the UK and the USA. Termed the 'New Right', this ideology challenged the consensus, widely held since the end of the Second World War, of the centrality of the state in the provision of welfare and other key services such as regeneration.

The position of the New Right ideology was cemented with the election victories of Margaret Thatcher's Conservative government in the UK in 1979 and Ronald Reagan's Republicans in the USA in 1980. The emphasis in urban regeneration turned towards the promotion of economic development through incentives to and partnerships with the private sector.

> Most significantly, during the 1980s there was a move away from the idea that the central state should or could provide all of the resources required in order to support policy intervention. This new policy stance was matched by a greater emphasis on the role of partnership. The more commercial style of urban redevelopment evident in the 1980s reflected yet another set of changes in the nature and structure of political philosophy and control.
>
> (Roberts 2000: 16)

The promotion of more commercial aims of urban regeneration through central government policy and in local authority practice has been characterised as a shift away from managerialism towards entrepreneurialism. While this characterisation does run the risk of ignoring some important continuities between the two periods, it is true to say that the 1980s did see local governments become more concerned with risk-taking and growth-oriented activities in the name of urban regeneration (Hall and Hubbard 1996, 1998).

> In Britain, urban policy underwent a radical transformation in 1979 informed by the philosophy of the New Right. This transformation has been described variously as a shift away from urban managerialism towards urban entrepreneurialism or privatism, or the shift from Keynesian to post-Keynesian policies.
>
> (Oatley 1998: 4)

This embracing of entrepreneurialism by local government was reflected in both the need to promote alternative forms of economic growth within their localities and to compensate for cut-backs in central government funding to local authorities, a reflection of the New Right view of the local state as an impediment to the free operation of the market. This period saw a widespread restructuring of the local state within the UK

Table 5.1 ***The evolution of urban regeneration***

Period *Policy type*	*1950s* *Reconstruction*	*1960s* *Revitalisation*	*1970s* *Renewal*	*1980s* *Redevelopment*	*1990s* *Regeneration*
Major strategy and orientation	Reconstruction and extension of older areas of towns and cities often based on a masterplan'; suburban growth	Continuation of 1950s theme; suburban and peripheral growth; some early attempts at rehabilitation	Focus on *in-situ* renewal and neighbourhood schemes; still development at periphery	Many major schemes of development and redevelopment; flagship projects; out-of-town projects	Move towards a more comprehensive form of policy and practice; more emphasis on integrated treatments
Key actors and stakeholders	National and local government; private sector developers and contractors	Move towards a greater balance between public and private sectors	Growing role of private sector and decentralisation in local government	Emphasis on private sector and special agencies; growth of partnerships	Partnership the dominant approach
Spatial level of activity	Emphasis on local and site levels	Regional level of activity emerged	Regional and local levels initially; later more local emphasis	In early 1980s focus on site; later emphasis on local level	Reintroduction of strategic perspective; growth of regional activity

Economic focus	Public sector investment with some private sector involvement	Continuing from 1950s with growing influence of private investment	Resource constraints in public sector and growth of private investment	Private sector dominant with selective public funds	Greater balance between public, private and voluntary funding
Social content	Improvement of housing and living standards	Social and welfare improvement	Community-based action and greater empowerment	Community self-help with very selective state support	Emphasis on the role of the community
Physical emphasis	Replacement of inner areas and peripheral development	Some continuation from 1950s with parallel rehabilitation of existing areas	More extensive renewal of older urban areas	Major schemes of replacement and new development; 'flagship schemes'	More modest than 1980s; heritage and retention
Environmental approach	Landscaping and some greening	Selective improvements	Environmental improvement with some innovations	Growth of concern for wider approach to environment	Introduction of broader idea of environmental sustainability

Source: After Stöhr (1989) and Lichfield (1992) in Roberts and Syjes (2000: 14).

that had a significant impact on the independence of local government and the extent of local democratic accountability. Both the independence of local government, which has traditionally been greater in the UK than in much of the rest of Europe, and the scope over which it had control were reduced. Control and provision of many activities previously delivered by local government were centralised or were the subject of control by non-elected local agencies or were privatised (Goodwin 1992; Ambrose 1994). Imrie *et al.* (1995: 32) have argued:

> [The] central state's restructuring of local government is reflected in the emergence of centrally appointed and directed agencies with responsibilities for many aspects of the governance of British cities. . . . They signify a new future for cities, including the removal of key powers from elected local government, the closure of their board meetings to members of the public and the press, the non-disclosure of the resultant minutes and records, and the pursuit of a politics of growth which seeks to enhance the powers of private sector interests in local economic development.

The model of regeneration which emerged during this period drew on developments that had taken place in a number of North American cities that had previously suffered mass de-industrialisation, the most notable example being Baltimore, which has pioneered the use of spectacular flagship projects of urban regeneration around its run-down Inner Harbour area since the 1950s (Bianchini *et al.* 1992: 246). The influence of American developments on British urban regeneration was readily apparent throughout the 1980s. There was a definite exchange of ideas on urban policy between the UK and the USA at the time (Hambleton 1995). The broad philosophy of urban policy in the two countries, as well as more specific characteristics, bore a close resemblance and the private sector was targeted as the main agent in the alleviation of urban problems.

Subsequent periods have seen both continuities and changes from the 1980s. The private sector is still an important partner in regeneration, although greater access has been granted to the community and voluntary sectors. The emphasis is now less firmly on the promotion of economic growth, although the enhanced global competitiveness of localities has been a priority of urban policy (Oatley 1998). Regeneration now tends to be viewed in broader terms to include social and environmental sustainability.

This discussion highlights the need to consider urban regeneration, not in isolation, but rather as being situated within a number of historical, geographical and political contexts.

Case study C

Regeneration and the legacy of the Athens Olympics

Festivals of various kinds, be they of culture, sport or a range of other events, are commonly used as a means of regenerating cities. The most highly prized festival, in terms of its prestige, international media exposure and potential for achieving regeneration, is the Olympics. Since the success of the 1992 Barcelona Olympics, these events have been used by cities to intervene in run-down urban districts and improve infrastructure, kick-start economic growth and enhance externally perceived images of places. Certainly much of the transformation of Barcelona, from a deprived, seamy port city into a successful, post-industrial metropolis, stemmed from the impacts of the 1992 Games on the city.

The 2004 Games were hosted by Athens, a city suffering from similar economic and social problems to Barcelona prior to the 1992 Games. The Athens Games were, like Barcelona's, an attempt to regenerate deprived parts of the city. For example, many Olympic facilities were located in deprived areas in the hope that they would lead to regeneration. However, few detailed plans were put into place to ensure that these goals would be achieved. Six months after the end of the Athens Games, there is little evidence that they have had any significant impacts upon the city's urban problems. Indeed, if anything the legacy of the Games appears to be problematic for the city rather than beneficial. Despite spending over £8 billion to stage the Games, the newly developed stadiums have largely stood empty since the closing ceremony and spending on the Games led to soaring public debt that looks set to continue. For example, the cost of merely maintaining the empty facilities is likely to be in excess of £50 million per year. While the Greek government has said that a number of domestic and foreign investors are interested in using the facilities, no firm details have been made available. Despite the success of the Athens Games themselves, the empty facilities that have been left behind have earned the nickname 'the herd of white elephants'.

Source: Howden (2005: 27)

Discussion topic

Urban regeneration is reflective of broader political ideologies. To what extent and in what ways do you think contemporary regeneration policies are reflective of prevailing political ideology?

Funding

The majority of funding for urban regeneration stems from central governments, although it is possible to find examples of local authority funds, charity funds, private sector funds and those stemming from the European Union. Whatever the origin of regeneration funding, funders have tended to use one of two models of allocation to decide where their money and resources should go. Until the 1990s the majority of regeneration funding was allocated on the basis of the demonstration of need, typically measured by looking at the scale and extent of social or economic depravation within localities. However, in the early 1990s, in the UK and elsewhere, the allocation of funding moved to a competitive bidding process. Under this allocation mechanism, funding was distributed on the quality of bids and the economic opportunities available in run-down localities rather than on the demonstration of need (Oatley 1998: 9). This represented an attempt by central government to use regeneration funds to foster innovation within deprived localities rather than to simply alleviate need. The shift to a competitive, 'Challenge Fund' model of funding allocation was a controversial one.

> The government and those who supported competition argued that it encouraged greater value for money and had a galvanising effect, motivating people to be innovative in developing proposals and encouraging more corporate and strategic approaches in regeneration activity. Critics of the Challenge model argued that it was a distraction, used to mask the decline of regeneration resources and mainstream expenditure and a way of rationing scarce resources. . . . Furthermore, competition between localities reduced the scope for inter-local competition and the new regime of governance tends to weaken local democratic accountability.
>
> (Oatley 1998: 9)

The Single Regeneration Budget, launched in the UK in November 1993, was one of the first sources of regeneration funding to be allocated using the Challenge Fund model. The result was a geographical pattern of allocation of funds that differed markedly from that of earlier regeneration rounds. It failed to include a number of severely deprived areas that had previously received regeneration funding (Shiner 1995).

The nature of policy and regeneration

Urban policy and regeneration initiatives can achieve their aims through interventions in a number of dimensions of localities. These include their natural environment, built environment, local social networks, economy, regulatory framework (including local planning regulations and taxes) and externally perceived image. It is common for regeneration initiatives to seek to address more than one of these dimensions. Having said this it is possible to chart broad shifts in the dimensions of localities that urban policy has sought to intervene in through time.

Until the late 1970s urban policy was overwhelmingly concerned with intervening in the built environment, particularly housing. This occurred either through the refurbishment of elements of existing housing stock, to bring them up to certain standards, or the provision of new housing. In the latter case this included provision of high-rise municipal housing, either replacing older housing stock in inner-city areas or in peripheral urban locations, or providing for expanding and increasingly suburban populations. This is not to say that this period was concerned solely with interventions in the built environment (see Table 5.1) but that it was the prime concern of regeneration at the time.

As the imperatives that urban regeneration sought to address expanded in the late 1970s and early 1980s, so did the aspects of localities within which it sought to intervene. Urban regeneration was still concerned with interventions into the built environment but in a very different way to previously. Urban regeneration also became concerned with regulatory framework of localities, their externally perceived images and, in a very limited way, their natural environment. The primary aim of much urban regeneration during the 1980s was improving the economic dynamism of de-industrialised localities by attracting private sector investments and relocations. Policies attempted this in a variety of ways. These included relaxing local planning restrictions and local tax burdens within specified locations (for example, in the case of the enterprise zone policy in the UK from 1981 onwards), the development of publicly subsidised flagship commercial developments, such as major office developments and convention centres, and addressing the negative images of many de-industrialised localities (for example, in the case of urban development corporations in the UK during the 1980s and early 1990s). Where elements of the natural environment acted as impediments to regeneration, for example in the case of contaminated land, a common problematic legacy of former industrial land uses in many cities, they were also the subject of intervention (Oatley 1993). Aesthetic improvements to the environment through landscaping were also a common component of urban regeneration programmes at the time. Indeed, the biannual Garden Festival programme in the UK was based entirely around this type of intervention in derelict urban sites.

Urban policy since the early 1990s has been less concerned with major interventions into the built environments of cities, although more modest interventions do continue; for example, the redevelopment of older urban landscapes to encourage the clustering of cultural and creative industries (Pacione 2001; see also Chapter 4, this volume). Rather the range of interventions has broadened. Recent interventions include enhanced engagement with the natural environment with the intention of promoting environmental sustainability. This follows the international adoption of Agenda 21 stemming from the 1992 UN Earth Summit in Rio de Janerio and its promotion through Local Agenda 21 (Selman 1996; see also Chapter 9, this volume). In addition, urban regeneration has sought to promote the enhancement of local 'capacity', defined as the structures and networks stretching between a range of stakeholders within localities. These include local authorities, the private sector, and voluntary and community stakeholders. In many cases the enhanced capacity of localities was deemed as valuable an outcome of regeneration as any material outcome. It was this enhanced capacity that was deemed likely to sustain regeneration beyond the life of specific projects. Policies such as the Single Regeneration Budget and City Challenge, launched in the UK in the early 1990s, and the USA's Empowerment Zone and Enterprise Communities programmes, from the same period, typify the more balanced, integrated approach to urban regeneration that has developed since the narrow, market-oriented, property-led approaches of the 1980s.

Discussion topic

Urban regeneration must choose to address either the basic needs of deprived populations or the need to make localities more competitive. To what extent can both of these aims be accommodated within urban regeneration?

Stakeholders

The range of stakeholders involved in or affected by urban regeneration varies from case to case. Similarly, the ways in which various stakeholders are involved in or affected by regeneration will vary both within and between cases. Healey *et al.* (2002; see also Pendlebury 2002) in a study of conservation and regeneration in the North of England recognised a wide range of stakeholders that they classified according to their relationship with regeneration. Those from the state sector affected include central and local government officers involved in a range of activities including traffic and street management, provision of housing, regulation, property development, conservation and business development. They also recognised a number of stakeholders with economic links with the area undergoing regeneration. These included providers of consumer services and product services, manufacturers, retailers, property developers, consultants and leisure suppliers. Finally they recognised a range of stakeholders from within civic society including residents, workers, shoppers, service users, leisure users and civic associations (Healey *et al.* 2002; Pendelbury 2002: 149). These stakeholders produced a complex web of linkages with the area undergoing regeneration that included local, regional, national and international links of various types. Clearly the stake that a resident, for example, of an area undergoing regeneration will have in that process will differ markedly from those of an international retail company with an outlet in the area, or an occasional user of leisure services who actually lives many miles away. Mapping the impacts of regeneration through such networks of stakeholders is a complex yet important task.

At the most general level it is possible to trace shifts in the relationships between key stakeholders over time. Those either most influential to urban regeneration programmes, or most affected by them, are central government, local government, the private sector, the community, and voluntary sectors and local residents. As the aims of urban regeneration became more economic throughout the 1980s business interests were granted greater access and power within urban regeneration. This was evident in the creation of central government agencies such as urban development corporations in the UK that not only took control away from local government but also passed it increasingly to the private sector through the membership of the non-elected boards which controlled these agencies (Imrie *et al.* 1995).

Since the 1980s policy-makers in the USA and the UK have increasingly advocated the partnership model of urban regeneration. Partnerships, in these cases, have attempted to include the community and voluntary sectors along with local authorities and the private sector. These represent attempts to broaden access to urban regeneration and return a degree of local accountability that was lost in policies such as the creation of urban development corporations. However, while this partnership or network regime model of governance appears to have emerged in the USA, the extent to which it is mirrored in the UK has been called into question. Davis (2003: 301), for example, has argued that central government has become increasingly influential within local policy, and, rather than developing or enabling local networks and partnerships to develop, has tended to retain a hierarchical structure of governance. In terms of the relationships between stakeholders, in UK urban regeneration at least, there appears to be a gap between rhetoric and reality.

All the examples of urban regeneration discussed in this chapter may be characterised as 'top-down' approaches in that they all emanate from central government. This tends to establish hierarchies of access and power and networks among stakeholders. However, since the 1960s and 1970s there have been numerous examples of quite different approaches to urban regeneration that have emanated from grass-roots sources. These have been termed 'bottom-up' approaches and have typically emanated from progressive local authorities or groups of residents, and offer alternative models of regeneration that engender very different sets of relationships among stakeholders. Examples have included progressive planning policies, community economic development and community architectural schemes (Pacione 2001). However, despite a number of examples where these initiatives have reaped important local benefits, in terms of the prevailing direction of urban policy and regeneration internationally, or on broad processes of urban development, their impacts have been marginal.

Case study D

Community regeneration in Bermondsey

The process of modifying derelict industrial units for commercial and residential use is one common to the inner areas of many former industrial cities. The units provide space for new activities entering these de-industrialised zones, often as a result of public sector subsidy, or provide housing for affluent city centre workers. This process represents what David Harvey (1989a) referred to as the insertion of one people's history by another. The former industrial workers are the ones who bear the brunt of de-industrialisation and tend to see themselves excluded from the future of their area. The result is often their physical displacement to make way for new forms of capital and investment and their associated populations.

An innovative artist-led regeneration is seeking to intervene in this process, however, in the south London borough of Bermondsey, using an old biscuit factory as the focus of the project. Former employees of the Peek Freen biscuit factory have been brought together by Paula Roush, an artist with Arts Council funding to explore the heritage of de-industrialised communities. The project, which seeks to explore the industrial history and identity of the area, aims to reclaim a past that is threatened with erasure by new rounds of investment and development. It is typical of many regeneration projects that originate outside the formal regeneration sectors and seek to challenge rather than affirm the new futures being mapped out for many de-industrialised communities. It is an example of regeneration offering a critical voice in the face of ongoing urban change that acknowledges little of the past of the areas in which it occurs.

While the project seeks to establish a permanent office in the old factory which is itself being rapidly converted into expensive flats, it is difficult to see how such projects can ever be anything other than peripheral to shaping the future of rapidly changing inner-city areas.

Source: Kelly (2004: 4)

Impacts of regeneration

After more than thirty years of formal urban policy and regeneration of towns and cities, it is apt to consider the extent to which these interventions have achieved their aims. What have been the impacts of urban policy and regeneration? To what extent have they had positive impacts upon cities, or on specific areas within cities? To what extent do pressing physical, environmental, economic, social and cultural issues remain?

To some extent, regeneration should be thought of as one strand in the ongoing process of urban change and development. It would be unfair to imagine that a successful regeneration programme would necessarily produce long-term, unchanging conditions of stability and prosperity for an area. Cities are dynamic and ever-changing. The external processes within which they are implicated are similarly dynamic. Consequently, once a regeneration programme or policy has run its course it is likely that cities will be facing new sets of local and global challenges. Having said this, it is possible to recognise both areas in which the impacts of urban policy and regeneration have been significant and also the persistence of long-term, seemingly deeply embedded urban problems where regeneration has, on a large scale at least, appeared to make much less of an impression. These and other aspects are considered in detail in Chapter 8. This section, however, outlines some key issues in brief.

Michael Carley, writing in 2000 about the impacts of urban policy on de-industrialised cities in the UK, argued that their regeneration had been 'partial'. A similar view could be taken of the regeneration of many urban areas in Europe, the USA and other parts of the world. The patterns described by Carley in these cities are ones that are common to many urban areas. The most visible and obvious impacts of urban regeneration have been in the centres of cities and around formerly run-down waterfronts and other 'heritage' districts. These were typically the foci of the major property-led regeneration programmes of the 1980s. These areas have benefited from major developments such as convention centres, hotels, major cultural facilities, office developments and luxury housing schemes. In terms of their physical appearance they have frequently undergone an almost complete transformation. The development of these 'people-attractors' has led to the growth of, often significant, visitor economies in these areas. There have been the obvious benefits of increased visitor spending entering the local economy and also less tangible positive impacts on city image and civic pride. At the same time, however, critics have raised doubts about the distribution of the benefits of regeneration among urban populations and about the nature of jobs created for disadvantaged populations (see Chapter 8).

However, Michael Carley and others have recognised the persistence of problems in many urban areas where the impacts of regeneration are much less obvious. While the transformation of many city centres and waterside areas is apparent, poor physical environments, characterised by derelict land and poor-quality housing, remain in other areas of cities. Similarly, many inner-city areas (and to this we might add peripheral housing estates) are characterised by the persistence of high levels of long-term unemployment, often passing from generation to generation, and attendant levels of poverty. The notion of 'urban renaissance', often justified on the basis of the transformation of city centres, would seem somewhat alien to many of the populations of these areas. Michael Carley has summed up the impacts of regeneration on British cities: 'while the nation has become better at property-led regeneration, it has not cracked the hard nut of helping households disadvantaged by long-term unemployment or the inability to work' (2000: 273).

Discussion topic

'Despite thirty years of regeneration, we have not cracked the problem of deprivation in neighbourhoods and cities hit hard by industrial jobs losses' (Carley 2000). Do you think this is an accurate assessment of the impact of urban regeneration over this period?

Conclusions

Urban regeneration initiatives have sought to ameliorate against the negative impacts of urban decline. They have attempted to achieve this either through the distribution of resources to disadvantaged populations or through the promotion of economic growth. Urban regeneration initiatives have tended to shift towards the latter approach through time. This was particularly apparent in the raft of market-oriented approaches that emerged in the 1980s. Recently the aims of urban regeneration have broadened to include, for example, the promotion of environmental sustainability. Despite a number of decades of urban regeneration programmes in the cities of Europe, the USA and many other parts of the world, and despite considerable improvements to the built environments and images of many central city areas, extensive, deep-seated problems of poverty and disadvantage remain. If urban regeneration is to be considered successful, these problems will have to be addressed and tackled, something that has seemingly been beyond many urban regeneration programmes so far.

Project idea

Look at a range of regeneration projects in a town or city with which you have been familiar over the past forty years. In what ways have the approaches to regeneration changed over this time? Use prompts from the framework for the analysis of urban regeneration to help you.

Essay titles

- In what ways do you think environmental concerns might be incorporated into urban regeneration projects in the future?
- Making cities competitive internationally is the main priority facing urban regeneration in the twenty-first century. How far do you agree with this statement?

Further reading

Gotham, K. F. (ed.) (2001) *Critical Perspectives on Urban Redevelopment*, Greenwich, CT: JAI Press. [A wide-ranging collection of critical essays on urban redevelopment.]

Imrie, R. and Thomas, H. (eds) (1999) *British Urban Policy: An Evaluation of the Urban Development Corporations*, London: Sage. [A collection of detailed evaluations of urban development corporations, the flagship urban regeneration programme in the UK in the 1980s.]

Imrie, R. and Raco, M. (eds) (2003) *Urban Renaissance? New Labour, Community and Urban Policy*, Bristol: Policy Press. [A collection discussing urban policy and regeneration in the UK since the election of the Labour Government in 1997.]

Pierson, J. and Smith, J. (eds) (2001) *Rebuilding Community: Policy and Practice in Urban Regeneration*, London: Palgrave. [An international review of urban regeneration policies and initiatives.]

Roberts, P. and Sykes, H. (eds) (2000) *Urban Regeneration: A Handbook*, London: Sage. [Provides an international overview of urban regeneration including key theoretical and practical issues.]

Web resources

British Urban Regeneration Association: www.bura.org.uk

Cyburbia. The Urban Planning Portal: www.cyburbia.org

Urban Regeneration Online Bibliography:
www.nottingham.ac.uk/sbe/planbiblios/bibs/urban/01.html

6 Transforming the image of the city

Five key ideas

- **All cities possess a range of partial and selective urban images.**
- **Place promotion, or city marketing, is an industry that has developed around the deliberate manipulation of urban image to promote economic development.**
- **Urban images assume ever greater significance within the post-industrial economy.**
- **The enhancement of image increasingly underpins the development of urban areas.**
- **Positive urban images have been said to be at odds with worsening economic and social realities in many urban areas.**

Introduction

The previous chapters explored some of the strategies that have commonly been employed since the early 1980s in the UK, mainland Europe and North America to regenerate the landscapes and economies of towns and cities. In this chapter the focus shifts to consider the transformation of the image of the city and the negative images that have become associated with many cities following the de-industrialisation of their economies, and which have proved a hurdle to their successful regeneration.

What is an urban image?

All cities have an image. In fact, it would be truer to say that all cities have, and always have had, a number of images. A place image of any kind is the simplified, generalised, often stereotypical, impression that people have of any place or area, in this case of cities. Yet it is impossible to know cities in their entirety. To make sense of our surroundings we reduce the complexity of reality to a few selective impressions. In being selective in this way we are producing a place image. Place images typically exaggerate certain features, be they physical, social, cultural, economic, political or some combination of these, while reducing or even excluding others. That the actual conditions in a city may have changed considerably since the image of that place was formed is not the point. In the world of perception the image is more important than the reality. This latter point may prove to be both an advantage and a disadvantage to cities. On the one hand, it means that an urban image can be cleverly manipulated and transformed by city marketers without the trouble of having to affect actual substantial change in that locality; on the other hand, it means that negative and increasingly misleading images may persist despite considerable change having taken place.

Forming urban images

Since the mid-1980s a large, and apparently growing, industry has developed around the deliberate manipulation and promotion of place images, which has become an integral part of urban regeneration programmes. This is explored later in this chapter. However, promotional campaigns by local authorities are not the only methods by which images of places are formed. Persuasive urban images may be formed in a variety of other ways. These involve:

- Media coverage of events in places which become the prevailing impressions of those places (e.g. the riots in various British inner cities during the 1980s and in Los Angeles in 1992).
- Satire and jokes about places, which tend to promote sterotypical views about them.
- Personal experience (e.g. visits to cities by tourists are frequently of short duration and by necessity highly selective, focusing on sites of interest or appeal and excluding large areas of cities).

- Hearsay and reputation (what people tell us about cities, whether from personal experience or hearsay, forms an important component of our impressions of places).

Bad urban images

Many cities have suffered the stigma of a bad image both now and in the past. One might think of the examples of Liverpool's reputation for crime, Manchester for drug problems, Los Angeles for social and ethnic conflict, and Birmingham (UK) for being an architectural and cultural wasteland. Clearly these are only partial truths in each case, since Liverpool does not have a monopoly on crime nor Birmingham on poor architecture. This is not the point. The images persist. Bad urban images tend to derive from the exaggeration of elements from a poor physical environment, a narrow and restricted cultural profile, social polarisation and unrest, economic dereliction and depression.

At this point it is important to say that urban images are not inherently good or bad in themselves. It is their relation to wider cultural fashions and trends that determines the ways in which they will be regarded. For example, during much of the twentieth century industry was a positive image for a city to possess. Changing fashions, linked to changes in the global economy discussed earlier, have caused industry to become regarded in a negative light. While industry was once equated with power, skill and pride, it is now more likely to be associated with dereliction, economic decline and pollution (Short *et al.* 1993). This change is reflected in the ways in which it has tended to be distanced from cities in their more recent promotional campaigns.

The rise of place promotion

The promotion of urban places has long been an integral aspect of urban development and hence urban geography. This is something that conventional accounts of the discipline have only recently recognised. Place promotion has affected the development of a wide range of urban environments in Britain, North America, Europe and Asia. It has involved a diversity of private and public institutions, the latter including all levels of local, regional and central government, and it has displayed distinctly different histories in different parts of the world (Ward 1994).

Some of the earliest examples of place promotion were found in the underpopulated west of North America from the early nineteenth century. Place promotion was used here to try to sell real estate at a time when many towns were being established (Holcomb 1990; Ward 1994). Place promotion in a form recognisable in the 1990s emerged in the north-eastern USA in the mid-nineteenth century and involved the promotion of industrial towns. Later in the century this promotion spread to the industrial provinces of Canada. Initially this type of promotion was largely a local activity which included institutions such as municipal boards of trade, councils and chambers of commerce, often in conjunction with private developers, railway companies and local fiscal initiatives. The main promotional publications to emerge from this were local newspapers, business and trade directories and promotional brochures. However, over the course of the twentieth century in Canada and the USA, place promotion began to involve progressively higher levels of state, provincial and federal government as it became integrated into regional development programmes (Ward 1994: 54–62).

Place promotion in Britain and Australia displayed very different histories from those of the USA and Canada, and focused, initially at least, on the promotion of different types of urban environment. Place promotion first emerged in Britain in the mid-nineteenth century with the advancement of mass tourist coastal resorts such as Blackpool and Scarborough (Ward 1988) initially by railway companies eager to drum up extra business, but later by the towns themselves. Such activity in the early twentieth century shifted to the promotion of the expanding residential suburbs, especially around London. Again railway companies were major agents of this promotion, but they were joined by private developers and building societies (Gold and Gold 1990, 1994). It was not until the 1930s that the promotion of industrial towns occurred on any great scale in Britain (Ward 1990, 1994). This was some fifty years after it had become widespread in the USA.

In Australia promotion was largely a post-war activity which formed a cornerstone of Australia's policy to attract immigrants and address its underpopulation problems. The main agent in this effort was national government, although local initiatives from individual towns and cities were also important (Ryan 1990; Teather 1991). Australia has also been marketed as a tourist destination by airline, travel and holiday companies and Australian tourist organisations.

Despite these early examples, place promotion has assumed far greater importance since the early 1970s, when waves of de-industrialisation

Table 6.1 ***Local authority promotional packages 1977 and 1992***

	% of local authorities	
	1977	*1992*
Guide	42.6	84.2
Glossy	29.7	56.2
Fact sheet	20.3	37.7
Industrial/commercial information	20.9	69.9
Tourist	28.4	84.9
Other	42.6	85.6
Slogan	43.9	45.2
Magazine/newspaper	—	32.2
Coat of arms	—	36.3
Logo	—	73.6

Source: Barke and Harrop (1994: 97)

generally affected the UK, Europe, North America, and in particular former industrial cities. First, more cities are recognising the need to promote positive images of themselves than was previously the case. It is now the exception rather than the rule to find an urban area not engaged in vigorous promotional activity of some kind (Table 6.1).

Second, the images produced by cities are more diverse than in previous years. Rather than simply promoting a unitary image of the city, industrial cities are recognising that they are not catering for a single, homogeneous audience, but for a plethora of distinctive niche markets. Consequently they tend to put forward a variety of images of themselves. Another aspect of this diversity is that a far wider range of organisations are involved in the business of place promotion. While previously a chamber of commerce, city corporation or local authority might be responsible for place promotion as well as a few private businesses such as railway and travel companies, the groups involved in the 1990s are more diverse. Some of these groups may be partly or wholly co-ordinated by a specialist national body (such as the Great British Cities Marketing Group) or local organisation (such as Birmingham's Marketing Birmingham); however, they may also include central government agencies such as urban development corporations, departments of the local authority, specialist facilities such as convention centres and airports, local institutions such as universities, which may have initiated schemes like science or research parks, members of the local business authority and the local media. The effect produced is a collage of often contrasting, but appealing, images of cities.

Finally, spending on place promotion, particularly as a proportion of local authority budgets, has increased since the early 1980s. This expenditure also formed a significant proportion of spending by urban development corporations and other agencies concerned with regeneration.

Despite these diverse histories the widespread and international process of de-industrialisation has seen the development of much homogeneity in the types of promotion undertaken by cities and consequently the types of image produced. Although this is not to say that alternatives do not exist, it is an inevitable reflection of the increasingly interconnected global economy.

Urban image in the post-industrial economy

A number of changes have occurred in the organisation of the global economy that have made the promotion of a positive image of place an extremely important part of economic regeneration. These were discussed in detail in Chapters 3 and 4. To summarise, it may be seen that the networks of competition which individual cities find themselves caught within have increased both spatially and numerically, and individual cities are subject to fewer protective measures and structures than has previously been the case.

In the UK, Europe and North America, many of the formerly most prosperous cities have suffered waves of de-industrialisation in their economies. They have broadly recognised that they might tap the growth displayed by the tertiary or service sector of the economy. However, these sectors are very different in their character from the secondary, manufacturing sector from which these cities' wealth was, in the large part, derived. The locational requirements of this sector included close proximity to raw materials, a large supply of labour, good transport links and accessible markets. In addition, a large amount of capital tended to be tied up in industrial plant. For profits to be generated this plant was required to engender economies of scale. These were achieved through large-scale production, usually of a uniform product for a long period. Consequently, this sector displayed a high degree of geographical *inertia*. They were locationally tied to specific types of site and, once established, needed to remain there for long periods to generate production economies of scale.

By contrast the growth areas of the economy now tend to display much less geographical inertia. The traditional locational requirements are now no longer as important as was previously the case. Cities therefore need to establish new advantages for themselves. Service sector activity involves little or no capital being tied up in heavy machinery or factories. The boom in the property development office rental sectors in the 1980s meant that there was a ready supply of appropriate space in almost every city. The growth in the development and employment of electronic communications technology has meant that markets can be instantly accessed from anywhere in the world. The traditional geographical attributes of location appear to be becoming an increasingly irrelevant aspect of the locational decision-making of service sector activity. Firms in the service sector are considerably more footloose than those in the secondary sector.

This has had important implications for the economic regeneration of cities. They have recognised that firms and investment now respond to a very different set of circumstances than previously. It appears that investment decisions in these growing sectors, given their different locational requirements, respond more to differences in image than to other more tangible locational factors.

> How an area is perceived and its physical or environmental desirability, however notional, will affect the levels of investment by industrial property developers, financial interests, and companies, on the one hand, and the inclination of employers and employees to work and live there, on the other.
>
> (Watson 1991: 63)

Footloose inward investment provides a potentially very unstable basis for urbanisation because it is able to switch location to a much greater degree than manufacturing or heavy industrial investment, to respond to even very slight changes in conditions elsewhere. This instability is exacerbated by three related conditions in the nature of urban economies and international economic systems. First, inter-urban networks of competition have become increasingly international in nature. Because of the global nature of many economic systems and markets (e.g. the business tourist and convention market), the choices of location available to decision-makers are potentially worldwide. Therefore, cities now commonly find themselves in competition with cities not only from their own country, but frequently from all over the world.

Second, not only has the spatial extent of networks of urban competition increased but so also has their numerical extent. Cities have traditionally displayed some degree of specialisation in their economic profile. While this certainly did not immunise them from competition, it at least tended to reduce the number of cities with which they were in competition. Networks of competition tended to develop among cities with similar economic profiles. However, cities found that these layers of economic specialisation were stripped away by the waves of de-industrialisation that affected them from the early 1970s onwards. The strategies adopted by cities to regenerate their landscapes and economies have tended to reduce this specialisation further. Urban regeneration projects in cities of the UK, Europe and North America have displayed a remarkable degree of similarity. The consequence of this is that cities have tended to find themselves in competition with an ever increasing number of cities entering the growth sectors of industrial and office relocation, business tourism and cultural tourism, as well as established cities that have traditionally dominated these sectors. This has raised problems not only at the local urban level but also at the national level. These problems have included those of market saturation and zero-sum growth on a regional or national scale, namely the problems associated with growth in one area resulting in decline in another.

Third, in many cases in the past, the economic position of many cities was secured, or at least protected, by protective structures or agreements. These included government regional policy, trade agreements between countries, military force or occupation and systems of empire. While these have far from vanished entirely from the world economy, they have certainly been reduced in number and scope and have changed in nature. Central government regional policy, for example, has been replaced by grants such as the European Commission's regional fund and City Challenge in the UK, which are decided by a process of competitive bidding between areas. These types of award have tended to increase the climate and extent of inter-urban competition and the insecurity of these types of regional investment.

The strategies of economic and urban regeneration employed by many formerly industrial cities in the UK, mainland Europe and North America appear to be caught within an increasing spiral of insecurity and competition. On the one hand, the markets they have and continue to enter have been characterised by an apparently increasing degree of competition. On the other hand, historically, whenever urban systems have displayed increased instability, the response of individual cities has been to increase their speculative urban regeneration programmes and associated promotional activities. This vicious circle of instability appears difficult to break. Certainly, the examples where cities have sought innovative responses to attempt to break it are rare. The majority simply fall back on to the, now rather unoriginal, routes of spectacular property development and urban promotion. While individual schemes may be judged as innovative, they are part of a wider strategy which forms an increasingly problematic basis for urbanisation.

Urbanisation and place promotion

The processes of place promotion are not incidental to those of urbanisation. As image assumes ever greater importance in the post-industrial economy it is becoming clearer that the actual production of urban landscapes reflects the necessity for cities to present positive images of themselves and that economic development is driven by programmes of place promotion. It is important to appreciate the distinction between 'selling' and 'marketing', since the two are very different processes. Selling is a process whereby consumers are persuaded that they want what one has to sell. However, marketing is a process whereby what one has to sell is shaped by some idea of what one thinks the consumer wants (Fretter 1993: 165; Holcomb 1993, 1994). The distinction between 'selling the city' and 'marketing the city' is therefore crucial to understanding their relationship with urban development. 'Selling' the city is likely to impact most directly upon the urban economy through increased visitor spending and outside investment. However, marketing cities also impacts directly upon their landscapes and development. It would be true to say that prior to the 1970s cities were largely 'sold'. It is now truer to say that they are 'marketed'. City landscapes are increasingly shaped according to views of what potential consumers want. Marketing the city, therefore, is a process that is increasingly integral to the shaping of urban development, rather than incidental to it.

> [Place marketing] is the principal driving force in urban economic development in the 1980s and will continue to be so in the next decade. . . . The logic that more jobs make a better city is giving way to the realisation that making a better city attracts more jobs.
>
> (Bailey 1989: 3)

> 'Marketing' is starting to replace the concept of merely 'selling'. . . . Selling is trying to get the customer to buy what you have, whereas marketing meets the needs of the customer profitably (or, in local government terms 'efficiently' at best value for money). Inevitably, this requires a much more sophisticated and more comprehensive approach affecting many local authority functions. Place marketing has thus become much more than merely selling the area to attract mobile companies or tourists. It can now be viewed as a fundamental part of planning, a fundamental part of guiding the development of places in a desired fashion.
>
> (Fretter 1993: 165)

Case study E

New images and urban development

Cities are being increasingly 'marketed' rather than merely 'sold'. Since the 1980s it has been true to say that urban landscapes have been shaped, in part, to appeal to potential consumers and have become more closely entwined in the processes of promoting cities. This is often referred to as 'investment marketing'. In 2003 the city of Birmingham (UK) underwent a rebranding exercise designed to change its image. The new image not only attempted to reflect the city at the time, it was also aspirational, in that it attempted to convey where the city wanted to be in the future. The promotion emphasised Birmingham as a diverse, multicultural, global city. Subsequent development and regeneration of the city sought to mould Birmingham in this new image. Clearly urban images and the processes of marketing are not ephemeral and unimportant. Rather, they are becoming increasingly central to the physical, economic and cultural development of places.

Plate 6.1 ***Enhancing urban image through new development, Bull Ring Development, Birmingham, UK***

The process of marketing cities

The ways in which the city has been promoted have been diverse. These have included distribution of guides, brochures and other information through tourist offices, libraries and commercial information services, through responses to postal enquiries, poster advertising, particularly in sites where large numbers of their intended audience are likely to gather (e.g. major railway stations and airports), through press advertisements, especially in the financial and property pages of the broadsheet newspapers, in specialist property or commercial pull-outs, and in specialist magazines, and through the employment of recognisable slogans and city logos. Increasingly cities are taking a proactive view of their marketing. Cities are now frequently sending out representatives, from the local authority and business community, to meet potential customers overseas. Throughout the late 1980s and early 1990s, prior to the opening of its International Convention Centre, Birmingham City Council ran a 'Birmingham roadshow' which toured Europe, and more distant markets like Japan, meeting potential customers and advertising the city through a series of face-to-face meetings.

The market for place promotion

Cities do not simply market themselves indiscriminately, but carefully target specific audiences, which include companies in the expanding service sectors of the economy and organisations involved in the planning of business tourist events such as conventions.

These audiences have an important determinate effect on the images that cities promote. The service sector does not tend to require a large unskilled labour force but a smaller, highly qualified and highly skilled labour force of largely middle-class professionals. Likewise business tourism usually involves middle- or high-ranking representatives of companies. The values that this audience is seen to respond to are highly specific and involve a concentration on lifestyle issues, including culture and environment, as well as business issues. While the markets may be broadly similar, cities are attuned to their subtle differences and direct their marketing accordingly.

New images for cities

A cursory survey of promotional literature from almost any town or city will reveal that they are keen to promote themselves as a good place to live as well as a good place to work. Cities emphasise not only their business opportunities but also their lifestyle activities. In the successful promotion of place the two are seen as indispensable and intricately intertwined. This section will explore in detail the images that cities have sought to create for themselves and promote as part of their economic regeneration.

Centrality

Now, as always, cities are desperate to create the impression that they lie at the centre of something or other. This idea of centrality may be locational, namely that a city lies at the geographical centre of England, the UK, Europe and so on. This draws on a deep-seated notion that geographical centrality makes a place more accessible, easing communication and communication costs. However, now that the economy is characterised more by the exchange of information than by hard goods, geographical centrality has been superseded by attempts to create a sense of cultural centrality.

Cultural centrality usually manifests itself as a cry that a city is at the centre of *the action*, meaning that the city has an abundance of cultural activities, such as bars, restaurants, night-clubs, theatre, ballet, music, sport and scenery. The suggestion is that people will want for nothing in this city.

Images of industry

Since the mid-1980s former industrial cities have been among the most vigorous place promoters. Their industrial history and the process of de-industrialisation have created something of an image problem for them. First, the most widely held images of industry are negative ones. They are of heavy, dirty, dangerous, polluting, rough, working-class activities. The process of de-industrialisation has created the images of dereliction, economic decay and unemployment – hardly the things to attract the well-heeled executive to a city. Promotional activity has concentrated on replacing these images of formerly industrial cities and changing the image of industrial activity generally.

> To call a city 'industrial' in the present period in the U.S. is to associate it with a set of negative images: declining economic base, pollution, a city on the downward slide. Cities with more positive imagery are associated with the post-industrial era, the future, the new, the clean, the high-tech, the economically upbeat and the socially progressive. We can identify a number of polarities in the division between industrial and post-industrial. Industrial cities are associated with the past and the old, work, pollution and the world of production. The post-industrial city, in contrast, is associated with the new, the future, the unpolluted, consumption and exchange, the worlds of leisure as opposed to work.
>
> (Short *et al.* 1993: 208)

The landscapes within which images of industry are framed are important in attempting to change the traditional images of industry. These landscapes are not those of the factory, the smokestack chimney and the storage yard, but are the highly designed green landscape of the science park. These typically contain well-maintained, landscaped lawns and gardens as well as lakes and sculptures. The buildings that populate this landscape are the postmodern, ornate, futuristic buildings of light, hi-tech, clean industry.

The promotion of actual industrial activity is similarly sanitised. The images of the production process contained in promotional activity specifically seek to change the image of industrial activity as heavy, dirty and dangerous. The overriding impressions created by these images are of technology, skill, precision and cleanliness.

Together these images aim to replace the traditional negative images of industry with more positive, appealing ones. These images suggest that industry represents the successful marriage of human skill, technology and environment, a far more attractive set of associations.

Industrial heritage sites have formed an important part of many cities' tourist strategies. This is so for a number of reasons, including the presence of large areas of prime central city property left empty following the decline of, for example, canal transport or dock activity. As well as restoration creating a series of desirable landscapes for the professional, industrial heritage may be used to generate positive images of places. This industrial tourism has been criticised due to the heavily idealised, sanitised treatment that it gives the past. Certainly, restored mills, docks, canals and heritage centres depict workers as cleaner, healthier and happier than they were ever likely to have been in reality.

The negative social attributes of the industrial past (e.g. the male dominance of employment and public life, or the role of ports in the slave-trade) are largely absent from modern representations. These images of industrial heritage play upon positive images such as pride, innovation, skill, strength and tradition.

Images of business

Dynamic images of places as good business locations have become a cornerstone of modern urban promotion. Rather than being organised around mundane functional attributes the promotion of locations for business has been dramatic and highly visual. It has focused primarily on images of architecture, communication and technology.

The spaces within which business is conducted, such as office towers, convention centres and business parks, have become crucial to the projection of city image and status. Buildings are important flagships of urban regeneration. Their size, design and capability are major icons of prestige. Consequently, the images of business play heavily upon the architectural setting, and in doing so perpetuate the role of the building and architecture in defining city status.

Increasingly, business is being conducted on a global scale. Electronic communications have allowed the instant exchange of information between vastly separated regions. This ability to conduct business globally has become another very important signifier of city prestige and status. Promotional images attempt to create the impression that their cities have the capability to annihilate the restrictions of space from a regional to a global scale, through physical transport (road, rail and air) and electronic communication. The prevailing images of cities as locations for business are of powerful nodes shaping a *global* economy, regardless of their actual position. These images are vital if cities are to attract investment and relocation from progressive businesses with international aspirations.

The suggestion of the presence of specialist and quality skills within localities is important, given that the labour requirements of business are ever more qualitative than quantitative. Consequently promotional materials emphasise the links between the industrial and business communities and the educational communities. These images act as signifiers of quality and excellence.

Case study F

Syracuse, New York: a tale of two logos

Syracuse is a city with a population of 147,300 (2000) and is located approximately half-way between New York City and Toronto. Its economy, which grew for almost a hundred years until the 1970s, was based initially on chemicals. From the early twentieth century it diversified, becoming a major world manufacturer of bicycles and typewriters. It was known as an 'industrial' city. However, since the 1960s the city has been affected by a slump in manufacturing which became particularly severe during the mid-1980s. Syracuse's original logo (Plate 6.2a), which dated from 1848, reflected both the dominance of industry within its own economy and the positive image associated with industry more generally. The old logo shows salt fields and smokestack chimneys. However, by the 1980s, industry had become associated with a whole range of negative images. In 1986 a competition was held to design a new civic logo. The winner not only reflected the shift in attitudes towards other aspects of the economy and the environment within Syracuse, but also, being designed to project a positive image to an external audience, reflected a wider shift in attitudes. Reflecting the rise of environmentalism within North American society, the new logo showed the newly valued aspects of Syracuse's environment, the recently cleaned-up lake and a modern, post-industrial skyline (Plate 6.2b). Excluding the industrial legacy of the city in representations such as this illustrates the shift of attitudes away from the industrial. Redesigning its logo in this way was one means by which Syracuse effected the transformation of its image from industrial to post-industrial.

Plate 6.2b ***Syracuse logo 1986***

Plate 6.2a ***Syracuse logo 1848***

Source: Short *et al.* (1993)

Discussion topic

Do you feel that the images promoted by cities are becoming increasingly alike and less and less original? Is this a symptom of the globalisation of the world economy and the increasing homogenisation of city life and landscapes?

Images of lifestyle

Urban promotion involves the selling of a location not only for business, but also as a place to live. It is vital, if people 'of the right sort' are to be attracted to an area, to suggest that it is equally able to offer lifestyle as well as business opportunities. These images of lifestyle tend to be predominantly anchored around two things: culture and environment. The use of leisure time is considered an increasingly important aspect of the decision-making process both for long-term relocation decisions and for short-term business (e.g. convention location) decisions and tourist decisions.

The notion of culture that cities have used within their promotional strategies has been very narrow and consists entirely of what might be termed 'high culture'. This typically consists of such pastimes as theatre, ballet, classical music, art galleries and museums. Where more popular activities are included they still reflect activities that are likely to appeal to a fairly narrowly defined target audience of affluent middle-class professionals. These include exclusive wine bars, restaurants and pubs, designer shopping facilities, cinemas and night-clubs. They suggest an abundance of exclusive or refined leisure facilities.

Images of environment

The essential complement to this cultural image is an equally appealing environmental image. Traditionally the environment of cities has been perceived very negatively. Since the mass industrialisation and urbanisation of the nineteenth century the image of the city as dirty, overcrowded, unfriendly and unhealthy has tended to dominate images of the city, in the UK and North America at least. Added to this, more recent images of economic recession, environmental decay and social unrest in inner cities and on peripheral housing estates have not helped to enhance the appeal of the city. Clearly cities still suffer from an image problem, some more so than others.

To counter this, local authorities have sought to re-invent or 're-imagine' the urban environment in their promotional campaigns. This has been achieved through the construction of a collage of positive environmental images, which typically consists of three elements – architecture, suburbs and countryside.

Architecture

Images of this tend to be of two kinds: spectacular and historic. Spectacular, futuristic, ornate, postmodern architecture such as flagship projects of urban regeneration suggests a progressive, dynamic city on the move. Images of historic architecture suggest civic tradition of some kind, such as local government (town halls) and art (galleries, concert halls).

Suburbs

These images are very domestic. They consist of prestigious, detached houses, manicured lawns, well-kept gardens, and suggest a 'nice', safe place to live.

Countryside

The presence of picturesque countryside near at hand is frequently emphasised in promotional materials. It offers an escape from the city, and somewhere to relax and for recreation (golf, sailing, climbing and walking).

Discussion topic

'The danger is clear: that city marketing strategies will deny any voice for, or celebration of, those elements of their indigenous culture that might impede the competitive search for wealth or jobs or a glossier image' (Barnett 1991: 171). Do you agree with the sentiment that city marketing is complicit in a process of social exclusion?

How successful is place promotion?

It is difficult to generalise about the success or the influence of place promotion campaigns in affecting relocation and investment decisions. The reactions of managers and decision-makers range from those who argue that advertising is incidental to their decisions, which are based on sound economic grounds, to those who argue that advertising is crucial to shaping the perceptions of their workforce and their clients. Despite the emphasis placed upon the process of place promotion by local authorities and other agencies, there is evidence that image is a less important factor in business relocation decision-making than is generally assumed. In a survey of companies relocated to Manchester, Young and Lever found that while the Central Manchester Development Corporation ranked the promotional image of the city as number one in a list of office location factors, this was ranked at only number nine by office managers of companies who had relocated to the area (1997: 337). Image was ranked by office managers below factors such as suitability/quality of office space, cost of office space and location with respect to customers. This suggests that promotional image per se is far less significant than promoters and local authorities appear to have assumed. However, this evidence does not signal the death of place promotion as a strategy for economic survival among former industrial cities. Rather, what it suggests is that cities are likely to modify their marketing strategies, perhaps marketing different images to local, regional, national and international markets (Young and Lever 1997), and adjusting the contents of their marketing, emphasising those factors that customers find important, over the promotion of glitzy images with little apparent content to back them up. Having cast doubt upon the significance of promotional image in certain markets however, it is likely that in other key markets for former industrial cities, most notably tourism and business tourism, promotional image is likely to be much more important. Where relocation is less than permanent, a matter of only a few days or so, image is still likely to hold the key to the decision-making process of potential customers. Despite doubts over its effectiveness, the sheer amount of place promotion occurring is likely to ensure its continuation. As competition between urban places for a whole range of public and private investment continues to intensify, it is likely that if places fail to advertise their claims, they will simply not get noticed.

Conclusions

> [City] councils which seek to promote a dream image of their city not only ignore the social consequences of focusing on private sector needs and perceptions: they can actually make matters worse for poor people.
>
> (Hambleton 1991: 6)

The images of the city created by promotional campaigns are highly selective. These images have been based around the emphasis of certain positive images and the exclusion of unappealing, negative images. It has been argued that this process is not innocent. It has been shown that it is an important part of the contemporary urbanisation process. It has been argued that in marketing cities towards a small, wealthy elite, the needs of less well-off people have been ignored. Some geographers have argued that the positive images of the city created in promotional campaigns and in urban regeneration have acted as 'masks' hiding the reality of urban problems (Harvey 1988: 35).

As with any process of urban change the benefits and problems are not evenly distributed. The following chapter considers how cities have redeveloped and regenerated their economies, landscapes and images to cope with the stresses of de-industrialisation. Chapter 8 considers the new social and cultural geographies etched out by this process.

Project idea

Select a series of historical and contemporary place promotion materials for a town or city with which you are familiar. To what extent and in what ways has the image of that town changed over time? Can you find reasons to explain the changes evident from your analysis of the materials? You may want to think of changes in fashions and tastes and how these might have affected the image or changes in attitudes towards industry, for example. The markets to which the town or city is promoting may also influence the images that are promoted.

Essay titles

- Place promotion involves the creation of fantasy images that divert attention away from a range of pressing urban problems. Discuss.
- The images promoted by cities are becoming increasingly aspirational, showing not only where they are at the moment but where they want to be in the future. To what extent are questions of image increasingly guiding the physical and cultural development of cities?

Further reading

Gold, J.R. and Ward, S.V. (eds) (1994) *Place Promotion: The Use of Publicity and Marketing to Sell Towns and Cities*, Chichester: Wiley. [A diverse collection of case studies.]

Holloway, L. and Hubbard, P. (2001) *People and Place: The Extraordinary Geographies of Everyday Life*, Harlow: Prentice-Hall (chapter 7, 'Representing place'). [Situates place promotion within a more general discussion of representation.]

Kotler, P., Asplund, C., Rein, I. and Haider, D.H. (1999) *Marketing Places in Europe: Attracting Investment, Industry and Tourism to Cities, States and Nations*, London: Financial Times/Prentice-Hall. [Explores the marketing of cities from a business perspective.]

Rose, G. (2001) *Visual Methodologies*, London: Sage. [An excellent discussion of a range of methods for analysing visual materials of all kinds.]

Ward, S.V. (1998) *Selling Places: The Marketing and Promotion of Towns and Cities, 1850–2000*, London: E and F.N. Spon. [An excellent history of place promotion.]

Web resources

Glasgow City Council Website: www.glasgow.gov.uk

Birmingham Online: www.birmingham.org.uk

Online Bibliography, City Marketing:
http://www.nottingham.ac.uk/sbe/planbiblios/bibs/urban/09a.html

7 Recent urban change

Five key ideas

- **There has been a broad shift towards postmodernism in many cities of the West (and in some cases beyond).**
- **The postmodernisation of the city is most apparent in the new architectural styles and spaces of city centres.**
- **Popular images of the inner city mask a more complex reality.**
- **Suburban areas have generally been subject to piecemeal, rather than fundamental, processes of change.**
- **Post-suburban developments are assuming much greater significance in the urban geographies of the twenty-first century.**

Introduction

A great deal has been written in recent years about the apparent transformation of the form and type of cities in Europe and North America. Much of this debate has focused on the emergence of 'postmodern', 'post-industrial' or 'post-Fordist' urban forms. Postmodern urban form is significantly different in its structure, its patterns of land values, its social geographies and its landscapes to the modern city described in models such as Burgess' concentric ring model (1925) and Hoyt's sector model (1933). Such cities, which developed over the course of the twentieth century, typically displayed homogeneous zones of land use and social group, with land values which declined regularly away from the centre of the city. Since the early 1980s, urban geographers have argued that this idea of the city is outdated and that we have been witnessing the emergence of new urban forms. Despite a number of differences between individuals they generally agree that these new cities are more fragmentary in their form, more chaotic in structure and are generated by different processes of urbanisation than earlier cities. This new urban form has been nicknamed the 'galactic metropolis' (Lewis 1983; Knox 1993). This describes a city which, rather than being a single coherent entity, consists of a number of large spectacular residential and commercial developments with plentiful environmentally and economically degraded space in between. They are said to resemble a pattern of stars floating in space rather than a unitary metropolitan development growing steadily outward from a single centre.

Some views of the postmodern city

The old idea of the city focused only on the picture postcard landmarks and the central crust of buildings and spaces. But it is clear that the present-day city has long-since outstripped those limits. The new incarnation of the city is an endless amorphous sprawl, with which outcrops of skyscrapers or vast shopping malls can appear almost anywhere.

(Sudjic 1993: 104)

Pushing one's face against the glass, one could see all that any human being could reasonably bear of St Louis: mile after mile of biscuit coloured housing projects, torn-up streets, blackened Victorian factories and the purplish, urban scar tissue of vacant lots and pits in the ground. It was the waste land. . . . Beside me the conventioneers were identifying another city altogether. They were pointing out the fine new home stadium of the Cardinals, Stouffer's Riverside Towers, the tall glass office blocks. To me the isolated sprouts of life in the surrounding blight were just objects of pathos: a few wan geraniums planted on a rubbish heap don't make a garden.

(Raban 1986: 329–30)

The modern, postmodern city debate

The debate about the apparent transition from modernity to postmodernity in society is extensive and has generated an enormous volume of literature from architecture and urban studies to film, literary criticism and fashion. This section aims to summarise the main characteristics of the debate that apply to the city. It should be used only as a crude, shorthand guide to the debate. The actual nature and extent to which modern forms of urbanisation have been supplanted by postmodern forms will vary enormously between cities. The outcome of the interrelationships between these two processes will be unique in every case. Not all cities, for example, may be said to be modern or industrial. Many, such as York and Durham in the UK, have retained much of their pre-industrial structure. Likewise, in the face of postmodern forms of urbanisation, many cities are likely to retain much of their modern or industrial structure. In reality many cities will demonstrate some combination of modern urban characteristics mixed in with newer postmodern urban forms. For example, recently redeveloped docklands areas may be surrounded by large areas of inner city barely affected by postmodern processes. In most cases the overall structure of the city still reflects modern urbanisation processes of industrial capital and planning. However, within this largely modern structure new urban forms have begun to emerge and there is evidence that the internal space of the city has begun to be resorted or reorganised (Cooke 1990: 341).

The main characteristics of the modern–postmodern debate with regard to the city are summarised in Table 7.1.

It should be apparent that these new urban forms do not simply appear for no reason; they are the visible outcome of a whole series of complex economic, political, social and cultural processes. It is vital that the urban landscape be read within this context. This is so for two reasons.

First, descriptions can provide only a very limited understanding of the urbanisation process. They say nothing of the underlying processes that created them. It is important to specifically examine these processes in conjunction with descriptive accounts.

Second, to say that the city reflects changes in these processes, while being true, reveals only one side of the relationship. Changes in international economics, politics, society and culture are not simply stamped upon the city without resistance. Rather, at the local level their imprint is uneven and contested. They are frequently opposed, resisted, misread or encouraged by institutions, agents and social groups in cities. The tensions between the operation of these processes and local groups result in tension that affects the outcomes of these often global processes at the local level. Therefore, the city not only reflects the nature of the processes of urbanisation, but also is active in affecting them.

Table 7.1 ***The modern–postmodern city: a summary of characteristics***

Modern	Postmodern
Urban structure	
Homogeneous functional zoning Dominant commercial centre Steady decline in land values away from centre	Chaotic multinodal structure Highly spectacular centres Large 'seas' of poverty Hi-tech corridors Post-suburban developments
Architecture, landscape	
Functional architecture Mass production of styles	Eclectic 'collage' of styles Spectacular Playful Ironic Use of heritage Produced for specialist markets
Urban government	
Managerial – redistribution of resources for social purposes Public provision of essential services	Entrepreneurial – use of resources to lure mobile international capital and investment Public and private sectors working in partnership Market provision of services
Economy	
Industrial Mass production Economies of scale Production-based	Service-sector based Flexible production aimed at niche markets Economies of scope Globalised Telecommuncations based Finance Consumption oriented Jobs in newly developed peripheral zones
Planning	
Cities planned in totality Space shaped for social ends	Spatial 'fragments' designed for aesthetic rather than social ends
Culture/society	
Class divisions Large degree of internal homogeneity within class groups	Highly fragmented Lifestyle divisions High degree of social polarisation Groups distinguished by their consumption patterns

The transformation of the city?

The debate outlined above is based primarily upon a small number of cities which have become constructed as archetypes of postmodern urbanisation. These cities, primarily, although not exclusively, in North America, have included Los Angeles, New York, Washington, DC, London and Tokyo. Aspects of this debate were discussed in detail in Chapter 2. Despite the undeniable and indeed growing influence of these cities both in their national economies and the international economy, they are not necessarily representative of the experience of urbanisation in the majority of cities in the West.

Despite obvious differences these cities all possess a number of overarching similarities. Most importantly, they are all world cities or global cities, control and command points of interlinked global economies and cultures (Hamnett 1995; Knox 1995). It is, consequently, these cities that are at the hub of emergent forms of urbanisation and it is not surprising that they have been drawn upon in the construction of new models of urban form. Despite words of caution from their authors, certain assumptions underpin both the construction and use of these models. The problems of applying these models, drawn from a select group of cities, are very similar to the problems in trying to apply the models of the industrial city which were based upon the 'shock cities' of the industrial revolution (Manchester and Chicago), to general urbanisation in the West during the twentieth century.

The assumption that the processes of urbanisation shaping cities like Los Angeles and New York in recent years will apply equally to other towns and cities in the West raises certain important questions as to the validity of these models. It is the aim of this chapter to evaluate the applicability of the model of urbanisation outlined above. In doing so it will examine urbanisation processes since the late 1970s. It is not the aim of this chapter to affirm or refute the model of urbanisation outlined above but to attempt a more subtle and hopefully meaningful critique. It will not ask: Should we apply this model of urbanisation but, rather: How should it be applied and in what ways can it be regarded as a blueprint of urban change in the twenty-first century? It will ask: Is it inevitable that all urban areas will be transformed in the ways that this model suggests? If this is not the case, why not? In what ways do the processes of urbanisation highlighted by the model (and discussed in other chapters of this book) feed into and change the urban hierarchy (the relations between cities) and the internal make-up of cities? What will be the spatial consequences of this?

This chapter will attempt to examine the effects of 'postmodern' urbanisation upon certain generic parts of the city. It will focus on recent urban change in the city centres, the 'inner' cities, the suburbs and 'post-suburban areas' primarily in the UK. In doing so it adopts the framework for understanding urban change outlined in the introductory chapter. It will examine the ways in which urban change has been produced, regulated and consumed since the late 1970s. This framework for exploring urban change is particularly well developed in the writings of the urban geographer Paul Knox (1991, 1992a, 1993) and is based on observation of the 'restless' landscape of and around Washington, DC. However, in situating these specific observations in a morc general context, he provides a useful insight into understanding processes of recent urban change elsewhere.

The production of urbanisation

The production of urbanisation has been affected by four main changes since the early 1980s. These have involved changes in the structure of investment in the built environment, the organisation both of the development industry and the practice of architecture and the technologies employed in building provision (Knox 1992a, 1993). Together, these changes have accounted for a number of emergent forms in the landscapes of cities.

Cycles of urban development closely reflect the waxing and waning of the property markets; on occasions they represent profitable destinations for investment, at other times they do not. Urban development is likely to be more intensive when the former rather than the latter is the case.

A major wave of urban redevelopment occurred between the mid-1970s and the property recession of the early 1990s. This period transformed a number of urban landscapes. This wave of development was sparked off initially by the effects of the oil price crisis of 1973. One effect of the rise of the price of oil was a massive influx of 'petrodollars' into the economies of the West. To employ a Marxist reading of the situation, over-investment in the primary sector made this an unprofitable destination for these funds and consequently they 'switched' to the more profitable secondary (property) sector. This switching was achieved through the actions of banks, investors and developers. A number of other sources of investment complemented the flow of petrodollars into the economies of the West. These included interest payments from Third World nations of their debts, investment from the growing economies of the Middle and Far East and investments from growing European pension funds (Harvey 1989b; Knox 1992a, 1993: 4–7). The symbolic (and very probably actual) end of this period of urban development was the bankruptcy in 1992 of Olympia and York, the developers of Canary Wharf in London's Docklands. The over-investment of the 1980s caused the collapse of the property market in the early 1990s.

While the pace of development was fuelled by the influx of investment, its shape, character and imprint on urban landscapes and urban form were determined by other factors. The development industry has undergone a considerable transformation in recent years, becoming both more concentrated and centralised into a smaller number of very large firms. Smaller firms have been eliminated to a large extent through competition and merger from larger firms. The average size of firm in the development industry has increased (Knox 1993: 5). The consequence of this for the urban landscape has been that these larger firms are able to undertake developments on a larger scale than was previously the case. This is evident in a number of new megastructures dotting the urban landscape (Crilley 1993; Knox 1993: 5–6). Increasingly in certain 'hot' sites, usually near the centres of world cities, property development has become a less local activity (Strassman 1988; Leyshon *et al.* 1990; Knox 1993: 6–7). The development of these sites has become increasingly bound up with the workings of the international economy. These megastructures have been developed mainly by large multinational developers, such as Olympia and York, who have relied on heavy borrowing from banks only too willing to lend to the booming property sector (Knox 1993; Sudjic 1993: 34). Formerly run-down regions have been favourite sites for such developments. Examples have included Battery Park in New York, California Plaza in Los Angeles and Docklands in the East End of London. It has been projects of this scale and staggering architectural style that have demonstrated the power of the large developer in reshaping or reorienting the landscapes of certain cities.

> While Canary Wharf was shaped by unabashed commercial opportunism, its closest contemporaries, the World Financial Center in New York, and California Plaza in Los Angeles, paid at least lip service to the strategic planning guidelines set for them by city governments. But planner or not, all these schemes served to demonstrate that it is the property developer, not the planner or the architect, who is principally responsible for the

> current incarnation of the western city. Large scale speculative developments – offices, shopping centres, hotels and luxury housing – shape the fabric of the present day city, not public housing and civic buildings. The developer, or more likely the institutions that fund his projects, pays the price for the land on which development will take place, determining far more rigidly than any zoning ordinance the range of activities that it may be used for. He [sic] chooses the architect and he sets the budget.
>
> (Sudjic 1993: 34–5)

Discussion topic

Sudjic (above) argues that commercial capital has become the major shaper of cities, relegating the significance of public buildings. Do you believe that public buildings are of little significance in shaping the forms and images of cities? What evidence can you find to support your point of view?

In the same way that other sectors of economic production have been able to become more flexible with the incorporation of computer technology and flexible labour practices into the production process, so has the production of the built environment (Knox 1991, 1993). This has mirrored the shift in production away from large-scale production dependent on economies of scale, towards flexible production aimed at the exploitation of highly profitable niches within the market, and generating profits through the range or scope of products available (Harvey 1989a). This emphasis on individuality and product differentiation is evident in new commercial, retail, leisure, industrial and residential landscapes (Knox 1993: 7–9; Graham and Marvin 1996).

This emphasis on product differentiation has had a reciprocally reinforcing impact upon the culture of the architectural profession. This profession has become increasingly divorced from the social ideals which guided its earlier development (see Jencks 1984; Harvey 1989a; Knox 1993). In previous periods, for example in the early post-war period, architecture, along with practices such as town planning, was suffused with social idealism. In the UK, for example, the design of the urban environment was seen as being important in both reflecting and helping to mould an egalitarian, democratic society, fit to accommodate the returning 'heroes' of war. It stood alongside the development of social welfare programmes of the welfare state and the National Health Service in the UK as one of the central canons of post-war society. Earlier architecture had regarded itself as a practice fashioning social utopias rather than just physical structures. This was the case both with the Swiss architect Le Corbusier's '*unit d'habitation*' and the earlier garden city movement.

The loss of architecture's social vision has coincided with its being co-opted by large institutional investors and speculative developers. This emphasis for the architectural profession has shifted to one of satisfying the demands of their clients, often engaged for a single project. This is typically achieved through decorative design and stylistic distinction, the massive scale of developments or the employment of a publicly known 'superstar' architect whose presence on a project will attract a great deal of media attention (Crilley 1993). Architecture, together with the architects responsible, has increasingly become a form of corporate advertising, and despite a few remaining radicals, has retreated from what it saw as its earlier social purpose. The consequence of this sea-change in the architectural profession has been that the urban landscape is designed in (artful) fragments and becomes littered with a number of 'spectacular', 'imageable' or 'scenographic' enclaves which are largely divorced from their immediate urban or social contexts (Harvey 1989a, 1989b; Crilley 1993; Knox 1993).

The regulation of urbanisation

The prime regulatory mechanism of urbanisation is the planning system. However, the culture of planning, very much like that of architecture, has also become increasingly divorced from its social origins. This has been so for a number of reasons, one being the failure of post-war planning to deliver the social utopias it had promised. This failure is best exemplified by the failure of the high-rise residential tower block in a number of British cities. These have become some of the most blighted residential environments and many face demolition. During the 1980s interventionist planning became regarded by the New Right as an impediment to the successful, free operation of the market. The New Right sought to reduce the influence of planning over the operation of the market as part of a sweeping reform of the public sector under the banner of 'freeing the market'. The planning system, through agencies such as the urban development corporations in the UK, became a mechanism for the facilitation, rather than the regulation, of urban and economic development.

This shift away from the regulatory role of planning helped usher in the fragmentary development of the urban environment. The planning system was able to conceive of the development of the city in its totality, and to situate developments within their wider urban context. Developers who have become increasingly influential agents of change in British cities since the 1980s, by contrast, have no need to look beyond the boundaries of their own development projects. They are concerned primarily only with the planning of their own 'fragments'. Planning in the 1980s and 1990s had become little more than the buttress of this fragmentary approach to urban development.

The consumption of the urban environment

None of the changes in the production and regulation of the built environment would have been likely to occur were it not for a concomitant change in the patterns and practices of the consumption of the urban environment. Put simply, new urban landscapes would not have appeared if no one was going to buy them. The relationship between supply and demand in this case is a complex one. It is difficult to say which had the major determinate influence; however, it is clear that they were reciprocally reinforcing.

It is true to say that contemporary urban culture is very consumption oriented. For those who are able to afford it, it acts as an important marker of status, distinction and identity. Consumption has always had this function to some extent; however, the patterns of consumption that arose during the 1980s were distinctly different from those prior to that time. The emphasis shifted towards notions of exclusivity, style and distinctiveness. Consumption at the top end of the market shifted away from the consumption of mass-produced goods, characteristic of the thirty years following the Second World War. Consumption patterns since the 1980s have fragmented into a series of niches determined by lifestyle or cultural preference.

Understanding the reasons behind this fragmentation of consumption patterns requires an appreciation of the psychological value of consumption. Consumption is a fundamental part of both individual and social identity construction. This applies equally to the consumption of places as well as commodities or ideas. Consequently the geographies of places of consumption (residential, commercial, retail and leisure) and the consumption of places is an important aspect of urban geography.

The new emphasis on consumption with urban culture derives from two phenomena, one social and one economic. First, the post-war 'baby-boomer' generation (born between 1945 and 1955) was characterised by an accent on the exploration of self-awareness. During the late 1960s this was achieved through the widespread adoption of a number of counter-cultural movements. However, the failure of these movements to deliver such self-awareness and freedom led to this urge being satisfied through the pursuit of individualised patterns of consumption (Knox 1991). This has become known as 'from hippie to yuppie' (Ley 1989). Second, the polarisation of economic opportunity and income, a pervasive economic characteristic noted elsewhere, created with them a number of jobs at the upper end of the spectrum that did not carry with them any inherent social status. This missing social status had to be constructed. Again consumption was the mechanism through which this was achieved (McCracken 1988; Knox 1991: 183–6, 1993: 19–25; Jackson and Holbrook 1995). As a result of this demand for status through consumption a number of key urban landscapes were reconfigured around conspicuous consumption. These were sites where consumption was conspicuous and conspicuous sites which were consumed.

Recent urban change

This section broadly surveys some recent changes in the forms and landscapes of cities over the past twenty-five years. It will examine in turn changes in the city centre, the inner city and the suburbs, and will conclude with a look at the emergence, or otherwise, of post-suburban developments in the UK, drawing comparisions with international developments.

The city centre

The centres of British and North American cities had by the early 1980s become much maligned places. Blighted by a legacy of inadequate post-war planning and design, they were typified by poor physical environments, pedestrian-unfriendly traffic systems, downgrading retail environments and economies dominated by offices that relegated the cities to cultural deserts after the early evening. Most importantly they were largely devoid of any economic mechanisms which could make positive impacts upon the local economy.

Since then, however, city centres have been the focus of an astounding degree of development. We have seen a widespread and comprehensive 're-imagination' of city centres (Bianchini and Schwengal 1991). This process has involved a combination of physical enhancement and cultural animation processes which should be viewed in parallel with the transformation of the images of cities which is described in Chapter 6. This re-imagination of the city centre is manifest in a number of landscape elements that emerged in city centres since the 1980s, and urban policies that aimed to animate these spaces through the promotion of an enhanced public or street cultural life.

Spectacular and flagship developments

Without a doubt, the most prominent landscape element to emerge in city centres since the 1980s was the spectacular or flagship development. These developments are of many kinds; however, what they have in common is their large scale and the emphasis on the importance of eye-catching, decorative, spectacular or innovative, typically postmodern, architecture. Flagship developments include office complexes such as Canada Tower at Canary Wharf in London's Docklands, convention centres such as the International Convention Centre in Birmingham and the Scottish Exhibition and Conference Centre in Glasgow, sports stadia such as the National Indoor Arena and the Don Valley Stadium near Sheffield, industrial parks such as the Dean Clough Industrial Park in Halifax, museums such as the National Museum of Film and Television in Bradford, and a whole range of temporary events such as sports tournaments, cultural, arts and garden festivals. These developments and events have frequently been initiated through some form of public–private partnership between local government or their agents and the private sector. Agencies such as the European Union have often been targeted as potential sources of funding. International examples of this include the redevelopment of Battery Park City in New York, La Defense project in Paris and various spectacular office projects in the previously neglected centres of American cities such as Los Angeles, Houston, Cleveland and Atlanta.

Such developments have both a tangible economic function as well as a less tangible, but no less important, symbolic function (Bianchini *et al.* 1992). These developments were intended primarily to act as catalysts, kick-starting the regeneration of the local urban environment and economy. Physically such developments are able to bring derelict land back into use and upgrade or enhance existing land uses. They also act as economic 'magnets' which are intended to attract people, spending and jobs, according to their type, in theory stimulating the wider urban economy (Harvey 1989b). They frequently act as stimulants to other economic development strategies by the local authority. It is from such 'ripple' effects that flagships derive their nickname. They not only attract or stimulate a 'flotilla' of smaller developments within cities but also may promote the development of policies or strategies that aim to spread the effects of development across the city (Bianchini *et al.* 1992).

Symbolically these flagship developments can act as central icons in the apparent transformation of a city's fortunes, image and identity (Bianchini *et al.* 1992: 250; Crilley 1993; Hubbard 1996). In the post-industrial economy, image is a vital component of economic regeneration (Watson 1991). Flagship developments can act both as important symbols of the rejuvenation of a city and as icons around which new images for that city might be constructed (Hubbard 1996). The 'scenographic' or 'cover shot' qualities of flagship developments are important in this respect (Crilley 1993; Knox 1993). The involvement of an enthusiastic local media, politicians and councillors is often vital in this effort (Thomas 1994). The hype that surrounds flagship developments is as vital a part of their function as any economic rationale behind their development.

Flagship developments of the types discussed here appeared initially in the USA. The first example was probably the Charles Center, a mixed-use development in Baltimore, which was built in the late 1950s. This was followed by a number of other spectacular developments around the city's Inner Harbour area (Bianchini *et al.* 1992: 246). The models of development pursued in Baltimore have proved hugely influential on those developed in the UK during the 1980s. The influence is apparent, for example, in the Broad Street Redevelopment Area in central Birmingham. This originally contained plans for 'Brindleyplace', a festival retail development based directly on Baltimore's Harbourplace development.

Flagship developments have been frequently sited within equally eye-catching 'packaged-landscapes' (Knox 1992a), which are designed in conjunction with flagship projects in order to complement the functions of the flagships. These landscapes include various permutations of architectural renovation, heritage themes, waterside developments, specially commissioned public art programmes and newly designed or renovated public squares.

The British fashion for flagship development was initiated primarily by the change in the political climate which forced local authorities to become more entrepreneurial and to fight to lure capital investment and jobs into their cities, rather than to expect financial help from central government. The flagship model of development was seen as a successful method of achieving these ends. Flagship redevelopment also reflected the postmodern conception of space inherent in the culture of much contemporary architectural practice and the planning profession. Flagships may be regarded very much as 'artful fragments' or 'scenographic enclaves' (Crilley 1993). They represent spaces which are shaped for aesthetic rather than social ends. There is little thought for how they might fit into the wider city economically, socially or culturally. Indeed, their failure to integrate into their surrounding social fabric gave rise to a number of descriptions such as 'yacht havens in a sea of despair' (Hudson 1989) which reflected the juxtaposition of conspicuous displays of wealth in the developments themselves and the often severe social and economic deprivation which surrounded them.

Festival retailing

City centres have traditionally been important retail foci. Although retailing displayed a trend towards decentralisation during the 1980s, certain parts of the city centre retail landscape have undergone considerable rejuvenation since the 1980s through the spread of festival shopping environments. This trend was closely associated with the spectacular redevelopment of the centres of British cities. Indeed, many festival shopping developments may be regarded as flagships in their own right. Again this followed earlier developments in the centres of North American cities.

Festival shopping can basically be regarded as up-scale shopping within themed environments, or as the product of the collision of two of the most important trends in urban culture since the 1980s: the physical revalorisation of previously redundant space and the construction of identity through consumption. Festival retailing may be thought of as the result of the shift towards consumption which highlighted the importance of exclusivity in the construction of identity. The recent retailing boom has been facilitated partly by the explosion of disposable income and the availability of personal credit among the young, affluent and middle class during this period.

The cultural shifts within which consumption was implicated made it far more than merely a functional fulfilment of need but a significant leisure activity in its own right. It became important, therefore, that its environments strove to convey an air of exclusivity. This was achieved through the use of ornate, postmodern design, often employing historical quotation and pastiche, through the theming of malls, or through claiming the cultural capital using buildings with historical associations and/or waterside locations. These developments often included restaurants and wine bars as well as shops. Festival retailing is a practice where the consumption of space and time is of equal cachet as the consumption of designer goods (Harvey 1989a).

Landscapes of heritage and nostalgia

The reclamation, renovation, reuse or reconstruction of past urban landscapes has become an almost ubiquitous aspect of the contemporary urban scene. These landscapes of heritage and nostalgia take on many forms. They include museums of urban and industrial history, the kitsch historical adornments of many packaged landscapes and new developments such as industrial hardware recycled as forms of street furniture, the renovation of old buildings or districts providing commercial, industrial, recreational or residential property and the do-it-yourself renovations of inner-city gentrifiers. Similarly the scale of these landscapes ranges from the minute adornments of historical signs or designs up to the thorough renovation of whole districts in or near city centres.

The popularity of heritage and nostalgic elements within new urban landscapes in the UK may be understood within the context of a national culture which, while having had a strong conservative and heritage tradition, has recently shown a renewed interest in nostalgic, idealised views of the past. This has been manifest in an explosion of heritage museums, nostalgic television programmes and films, writing, advertisements and other media. Many commentators have interpreted this as a retreat from the dreadful conditions that prevailed in modern Britain since the mid-1970s (Hewison 1987). However, an alternative and more positive interpretation of the heritage impulse has argued that this interpretation is overly elitist and dismisses the validity of popular conceptions of history. Critics of the earlier interpretation have argued that the heritage industry offers a more democratic and accessible experience of history (Samuel 1994).

In an international context the incorporation of past landscapes into the contemporary urban scene also fits with the postmodern turn in architecture and urban design. Namely it is able to satisfy the postmodern yearning for eclecticism, the vernacular, regional distinctiveness and decoration through historical quotation. Heritage, or rather the very obvious possession of a past by an urban space, expressed through either a famous past or prestigious historical architecture has proved a valuable commodity of great appeal to consumers of urban space since the 1980s and 1990s.

Cultural animation

While the physical changes described above might provide the template for the re-imagination of a 'new' city, physical change is not all that is required to regenerate city centres. It became obvious during the 1980s that public culture in many city centres was very impoverished. The centres of large cities in the UK and North America provided little more than spaces for working and shopping. An active public cultural life was regarded as a necessity for successful regeneration of urban space. By contrast many European cities were known for their vibrant, lively public culture. Where spaces for such public culture existed in British and North American cities, by the early 1980s they were downgrading, a legacy of their generally poor planning, or they had been appropriated, in popular perception at least, by antisocial behaviour. The rejuvenation of public urban culture in these cities in the 1980s was achieved through the physical improvement of city centre public space and a range of local authority policies which aimed to improve the safety and accessibility of city centres and to encourage open-air, free events. Policies have included street lighting, pedestrianisation, relaxed licensing hours, the introduction of closed-circuit television, traffic calming and improved public transport (Bianchini and Schwengal 1991).

The provision of new or redesigned public squares, often complete with extensive programmes of specially commissioned public art and entertainment such as street theatre and musical displays, has been a common response by city councils. Birmingham replanned two major civic squares in its centre during the 1990s. Both had large budgets for the provision of public art. Similarly in Sheffield, Tudor Square has been transformed into the setting for a number of free public events.

The shifts in design and policy outlined above reflect a recognition of the importance of the 'night-time' economy, both to the urban economy and to the enhancement of the cultural life of the city (Montgomery 1994). This was a very deliberate attempt to reorient urban culture more along European lines than had previously been the case.

Discussion topic

It is an oft expressed view that we are currenlty experiencing a renaissance of the city centre. What evidence can you find to support this? Is this supposed renaissance of the city centre universal or does this view hide a more complex reality? For example, are all town and city centres with which you are familiar enjoying a renaissance? Is everyone able to share in the benefits of this renaissance?

Fortress landscapes

Mike Davis' seminal (1990) *City of Quartz: Excavating the Future in Los Angeles*, a book about the brutalisation of urban life in Los Angeles and the repression of its poor and ethnic minorities, reads like a war correspondent writing from the front line. Davis described in impassioned prose the multitude of ways in which the restructuring of Los Angeles in the 1980s and 1990s was driven by a paranoia among the middle classes and the rich of perceived threats from a variety of 'others' such as black gang members, drug users and the homeless. The consequences of this were the production of a carceral city that hived off the rich in heavily protected enclaves and excluded others. Davis' book was one of the first to recognise an increasingly prevalent aspect of the urban landscape: fortress landscapes.

Fortress landscape is something of a catch-all term for landscapes that are designed around security, protection, surveillance and exclusion. Other terms that have been applied have included landscapes of security, defence or paranoia. While the fortification of the urban landscape in various ways has long been a feature of cities, for example for military purposes, Davis argues that it is now a defining characteristic of postmodern urbanisation.

> This obsession with physical security systems, and, collaterally, with the architectural policing of social boundaries, has become a zeitgeist of

> urban restructuring, a master narrative in the emerging built environment of the 1990s.
>
> (Davis 2003: 202)

The examples that Davis cites include the use of gating around suburban communities, surveillance systems in commercial and residential areas, the prevalence of armed security, the denudation of public space, public life and amenities, and the design of pedestrian and traffic flow systems to reduce or prevent contact between newly designed and older, poorer, areas. In addition, he, and others, have recognised the use of architectural symbols and public art to lend areas particular cultural 'auras', marking them out as areas for a narrow, socially homogeneous public (Miles 1997; Hall 2004).

His survey of Los Angeles in the 1980s and 1990s, prophetic in many ways, makes a number of points about the emergence of fortress landscapes in the city. Much of the fear that drives the urban restructuring, he argues, is perceived rather than actual. This is typically driven by racist media and politicians who portray certain groups as a threat to security and order, generating moral panics among urban populations. The invention of perceived threats, against which politicians can then appear to act, again with the complicity of a sensationalist media, is a common tactic in urban and national governance. Further, Davis recognises that fortress landscapes have both internal and external effects. Internally, they act to ease the paranoia of those whom they enclose and protect. However, conversely, in reducing interaction between socially heterogeneous groups, who now encounter each other only through the distorting lens of the media, they perpetuate it and concomitantly the need for security. Externally, fortress landscapes have emerged as a major dimension of the exclusion and repression of large sections of the urban population who are hived off in desperate inner-city barrios and ghettos. Finally, Davis argues that the paranoia and the desire for security and exclusion of 'others' that underpins the restructuring of the urban landscape has become deeply entwined with urban policy in Los Angeles and beyond. Davis (2003: 204) sums up the situation thus:

> Unsurprisingly, as in other American cities, municipal policy has taken its lead from the security offensive and the middle-class demand for increased spatial and social insulation. De facto disinvestment in traditional public space and recreation has supported the shift of fiscal resources to corporate-defined redevelopment priorities.

As with so many aspects of urban geography, Los Angeles is the most extreme example of this process of urban restructuring. When Davis first talked about the militarization of urban life in Los Angeles in the early 1990s few of his observations would have been recognisable to residents of all but a handful of major cities in the USA and beyond. His writing appeared to belong to the realm of science fiction, painting a distopian future rather than reporting from an actual present. However, since *City of Quartz* first appeared in 1990 the processes that Davis observed have both become more deeply embedded in the urban geography of Los Angeles and have developed in many other cities around the world, albeit rarely to the extreme degrees that they are present in Los Angeles. It is now possible to recognise fortress landscapes as a characteristic of the urban geographies of many cities. For example, the use of impenetrable facades, gating and the electronic control of access is an increasingly common element of newly built luxury residential developments, especially in the case of brownfield developments built near the centres of cities or in inner-city areas (see Plate 7.1) and surveillance and CCTV

Plate 7.1
Security and luxury, inner city residential developments, Birmingham, UK

cameras are an almost ubiquitous feature of European city centres and are increasingly common in residential areas (Speake and Fox 2002: 15). It seems that security, if not necessarily paranoia, is becoming an increasingly significant driving force underpinning change in the urban landscape.

The inner city

In debates about the form and character of the postmodern city, the inner city is conventionally painted as the 'sea of despair'. It is portrayed as a zone abandoned by both formal economic mechanisms and conventional forms of social control and regulation. These, it is argued, are replaced by forms of twilight or informal economy and social control based upon violence or threat. These images first emerged in coverage of social and racial disturbances in areas such as South Central Los Angeles. These images formed the staple coverage of a whole variety of media, both fictional and non-fictional (including some academic coverage). Such nightmare images of the inner city have, since the early 1980s, begun to appear in accounts of the inner areas of a number of British cities. The inner city has come to represent the dark side of the postmodern city, sharply juxtaposed with the spectacular redevelopment of the city centres. However, it is important not to let the moral panics stirred up by such representations cloud the reality of life in inner cities.

The inner city possesses both a formal geographical identity and a metaphorical identity. Geographically the inner city is that area between the city centre and the suburbs traditionally referred to in the geographical models of Burgess and Hoyt as the 'twilight zone' or the 'zone in transition'. It was an area of high-density housing, largely terraced, which dated back to the late nineteenth century, mixed in with local authority housing redevelopments. It grew up around nearby heavy industry before it was closed down or relocated to the suburbs, creating another set of problems for the inner city. The inner city was associated primarily with working-class and immigrant populations. However, this strictly geographical definition has become somewhat surpassed by its metaphorical identity. The rigidly delimited zoning and internal homogeneity of the urban form of the industrial city has been dissolved to some extent recently. As a result of this the term 'inner city' has now come to refer less to a specific geographical location and more to a set of social, economic and environmental problems perceived (often incorrectly) as being typical of the inner city. Such problems, however, are as likely to occur on municipal housing estates on the edges of cities as they are in the geographically inner areas of cities.

The inner city has always been regarded as the lower-class environment of the city. What is distinct about recent commentary, however, is the equation of the inner city with the existence of an urban underclass. The inner city has become regarded not as the poor relation of the city, but as a member of a different family altogether. This echoes wider debates about the polarisation of the economy and the severing of the links between groups at the lower ends of the spectrum and those at the top. The inner city has become the imagined nightmare environment of the disenfranchised and the excluded.

Such moral panics of urban spatial and social forms have begun to be transferred into new urban forms such as recently designed suburban settlements for the affluent middle classes in cities such as Washington, DC and Los Angeles. These settlements, such as Park La Brea and CityWalk in Los Angeles, display a desire among the middle classes for surveillance of their property, protection and security (Beckett 1994). This 'citadel' or 'paranoid' architecture is in response to a perception of the poor as 'other', a threat (Soja 1996: 204–11). These panics are whipped up by selective coverage of these areas in the media, coverage which focuses only on the newsworthy aspects of these areas, namely crime, violence, drugs, gangs and apparent social breakdown.

This perception of the inner city is reflected in the geographies of service provision and abandonment. The inner city emerges constantly as the area most qualitatively and quantitatively under-provided with basic health, welfare, education and financial services. Banks, for example, appear increasingly to be adopting a policy of 'cherry picking' their customers on the basis of income. This inevitably has a geographical expression in terms of access to financial services. The policy is expressed in two ways. First, the geography of financial service provision is changing, with banks closing outlets in inner-city areas and shifting provision towards areas with more affluent social profiles (Leyshon and Thrift 1994). Second, an increasing proportion of bank business is transacted electronically. In 1988 Midland Bank set up First Direct, the UK's first 'electronic bank', which uses electronic telecommunication systems to deliver all of its services and has no physical branches on the high street, only one single central office. The bank's customers are chosen specifically on the basis of their incomes, focusing on those earning an income above a certain level (Graham and Marvin 1996:

149–50). The distribution of these customers, therefore, reflects that of the traditional banking outlets, concentrating in areas of affluence over areas of poverty. This process is referred to as 'financial exclusion' (Leyshon and Thrift 1994).

Such attitudes towards the inner city are not new. In many ways the contemporary representation of the inner city reflects very closely those of middle-class Victorians both in Britain and North America. The rapid and largely unregulated growth of the Victorian city generated a series of moral panics centred on fears of crime, social disorder and disease (Coleman 1973; Mayne 1993). These were a staple of satirists, journalists and social reformers at the time.

Returning to the more strictly geographical identity of the inner city, it is certainly apparent that examples of the 'seas of despair' discussed above do exist. However, this image should be qualified. The inner city is a far more multifaceted entity than this image would suggest. First, simply to regard all inner cities as constituting seas of despair is a vast oversimplification. It hides far more than it reveals. Second, it is worthwhile asking if the conditions existing in inner cities are worsening. Are inner cities progressing along a common trajectory towards becoming seas of despair? While there is evidence that in general the problems typically associated with the inner city are worsening, to label this trend universal would be to oversimplify a very complex series of trajectories. This label ignores the fact that while the economic factors affecting conditions in inner-city areas may be general, they are mediated locally. The outcome of this interaction of the general and the local is a complex mosaic of local difference. The local economic, political, social and cultural conditions of inner-city areas are as crucial to highlight as the general economic processes with which they interact. Consequently, for every inner-city area or peripheral council estate apparently in the grip of economic decay and social dislocation and breakdown, there are many others where, despite conditions being worse relative to other areas in the city, the reality does not fit the prevailing media representation. Even within areas notorious for economic, social and environmental problems, such conditions are rarely universal. Such areas, despite media representations to the contrary, are internally heterogeneous. Conditions within them are far from universally dreadful.

Media portrayals of the inner areas of cities are selective and partial. That which is deemed newsworthy is by definition exceptional. The ordinary and the everyday simply do not get reported in newspapers and on television. There is a danger that in examining only the exceptional events media representations are mistaken for reality. This stigmatises inner-city areas and peripheral housing estates. It may be that conditions and opportunities in urban areas in general are becoming more polarised, but at a local level this process crystallises out as a mosaic of local complexity which belies such simplistic images as the 'sea of despair'. Inner-city areas should be recognised as complex; however, providing a detailed guide to such complexity is beyond this short section. Rather, it has tried to provide a framework within which the character of inner-city areas can be appreciated.

Gentrification of the inner city

Gentrification is a term that has come to refer to the movement of affluent, usually young, middle-class residents into run-down inner-city areas. The effect is that these areas become socially, economically and environmentally upgraded. Gentrification is a process that has generated ongoing geographical debate and dispute since the early 1980s.

It is generally argued that these middle-class residents, drawn by the appeal of the 'buzz' of inner-city living, are attracted to the cultural capital attached to distinctive architecture, often available at very low prices due to its physical dilapidation (Fincher 1992). Such a view, however, disguises considerable variations in the nature, extent and impact of gentrification. While gentrification is an established and extensive practice in many North American, Australian and some European cities, in the UK, with the exception of London, it is less apparent. Similarly there is a wide variation in the groups involved in the gentrification of the inner city. As well as the young, middle-class professionals, artists seeking cheap, extensive work space (Zukin 1988) and women seeking access to city centre employment opportunities and good public transport have been identified as gentrifiers (Rose 1989; Fincher 1992: 107). Local authorities have promoted a number of activities that support gentrification, such as the refurbishment of old industrial space to provide studios for artists and cultural industries, as a way of economically revitalising and diversifying inner-city areas. One well-appreciated aspect of gentrification is the potential impact it has upon existing working-class residents. These impacts have included social disharmony as new groups enter community space and the displacement of working-class groups as house prices rise in the wake of gentrification. These problems have been very apparent in the urban development corporation-led gentrification of London's Docklands (Rose 1992) and the gentrification of the Lower East Side in New York, led by the real-estate industry and encouraged by the City (Reid and Smith 1993). Clearly the extent, diversity, institutional linkages and impacts surrounding gentrification should not be overlooked (Smith and Williams 1986; Hamnett 1991).

The suburbs

The suburbs may be defined as the outer areas of a city which are linked to thc city by their lying within the commuter zone of an urban area. They usually form a continuous built-up area. 'Suburbs' usually refers to the predominantly residential landscapes built up around the urban core as a city has expanded outwards. Suburbanisation of the residential population of cities is essentially a twentieth-century process closely associated with development of transport technology: the trolley-bus, the train and the car respectively. The suburb has become the landscape of the middle class and the skilled working class internationally.

Despite being burdened with an image that suggests monotonous regularity, the suburban areas of cities, or indeed of any single city, display a great variety. This variety derives in large part from the different periods during which suburbs have developed (Whitehand 1994), the markets for which they developed (Knox 1991, 1992a, 1993), their physical settings, the architects and developers employed, and the local planning framework and operation of the system (Whitehand 1990). Again in contrast to their image of unchanging regularity, the suburbs are dynamic landscapes which demonstrate complex processes of change (Moudon 1992; Whitehand *et al.* 1992).

Suburban development and change

Over the course of the twentieth century, suburbs have tended to expand outwards through addition. A number of phases of growth can usually be recognised within cities which relate to a number of social, economic and technological factors. These have given rise to a series of distinctive landscapes. In the UK these phases include Victorian, inter-war, post-war and recent development. Each is distinct in the landscape it produces.

Although these phases of development have been successively outwards, they have not been uniformly smooth. The pace and style of suburban development is related very closely to booms and slumps in house-building cycles. Slumps in house building, related, for example, to credit availability, interest rates on borrowing and demand for housing, generally cause land values to fall. This has a number of effects on the amount of house building occurring, the size of houses and plots, and other types of building and land use included in suburban areas. During slumps it becomes feasible to build more houses at lower densities on larger plots of land and to build more non-residential buildings such as institutional or commercial buildings, and to include extensive land uses such as sports pitches and playing fields. The characteristic landscapes formed during building slumps are referred to as fringe-belts. During building booms, outward development tends to be more rapid, housing densities higher and non-residential land uses fewer (Whitehand 1994).

Other important influences upon the form of suburban development are innovations in building and transport technologies and the pace at which these are adopted. Evidence has shown that there has been a particularly close association between the adoption of various innovations in transport and bursts of house-building activity, again with each boom producing a characteristic landscape.

A model of suburban development has been proposed which includes each of these factors of change: additive growth, fluctuations in building cycles, and building and transport innovations (Figure 7.1).

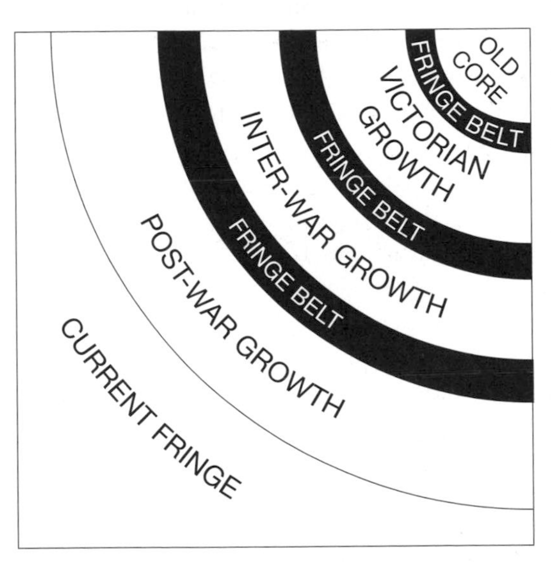

Figure 7.1 ***A model of suburban development***

Source: Whitehand (1994: 11)

Recent suburban change

In contrast to the more dynamic city centres, the suburbs of British cities have recently been subject to changes that have been more piecemeal than epochal. Once built, suburban residential zones are far from static, and display change throughout their lives. The suburban landscapes of Britain have changed predominantly through processes of addition, replacement, refurbishment, conversion and conservation since the mid-1970s. One of the most important changes in the British suburban landscape has been the conversion of single-unit dwellings into multi-unit dwellings such as flats and the infilling of vacant land in large plots, usually the gardens of large detached houses. Although these have been widespread processes across the UK, they have demonstrated some regional differences, with the South East region under greatest pressure to increase dwelling densities in its suburban areas (Whitehand 1994). This process is in sharp contrast to the rapid and extensive outward expansion of suburban areas in the first half of the twentieth century which was fuelled by, and in turn fuelled, the rapid spread of the motorcar.

The redevelopment of Britain's existing suburbs and the development of new residential areas has become a major planning issue in the early part of the twenty-first century. This raises the obvious question: Where will these homes go? Not only is it unlikely that the existing housing stock will adequately provide for this increased demand in terms of sheer numbers, but it is also probable that the existing housing stock will not provide houses of an adequate type for this new demand. The outward expansion of British suburbia proceeded on the assumption of an 'ideal' household which consisted of a nuclear family. However, such an ideal type is increasingly not the case in the UK, which has seen an increase in alternative types of households. These have included single-headed households, extended families, couples delaying marriage and children, and an increasing number of people choosing to live alone. This demographic shift has led to a demand for several different types of housing. It is likely that new dwellings of different types will have to be built to cater for this new demand and that much of this new building will be focused on smaller urban settlements. The possibility of building new towns or settlements on greenfield sites has also received renewed interest.

The outward expansion of towns and cities has been restricted through the introduction of restrictive planning policies such as the 'green-belt' policy in 1938. The green-belt policy, and its variations, has been designed to ensure the protection of vacant green land between cities. The policy was intended to prevent the unchecked urban sprawl swallowing up the countryside between cities. Despite coming under considerable pressure since the late 1970s it remains a powerful impediment to urban growth.

The meanings of suburbia

As well as possessing a very distinctive physical identity, British suburbia possesses an equally distinctive cultural 'identity'. This is evidence of the fact that it is the repository of a number of important cultural values. English culture has long demonstrated a strong anti-urban streak which may be traced back to the mid-nineteenth century. By contrast, it has also displayed a powerful veneration of an idealised 'chocolate-box' vision of rural England. The suburbs may be read as a response to the need to live near the centre of cities while wishing to attain an essence of rurality. Despite being essentially part of the urban settlement, suburban areas have long been imbued with associations of the rural, through their physical design, which has been centred on the garden, their naming and their promotion and representation (Gold and Gold 1994).

The continued manipulation of this image has ensured that suburban areas have become imbued with values long attached by the English middle classes to the rural idyll: tranquillity, peace, community, safety, an unhurried pace of life and domesticity (Eyles 1987). This identity has become the focus of sharp criticism by feminist geographers.

Geographical perspectives on the suburbs

The suburbs have been subject to a number of different interpretations by geographers adopting contrasting perspectives. While they are not competing interpretations, because they are concerned with different aspects, they emphasise the range of economic, social, psychological and cultural perspectives which geographers have adopted to study the urban and the suburban.

Marxist geographers have interpreted the massive suburban expansion of the cities of the USA, Canada, Australia and the UK in the inter-war and post-war years as an attempt to resolve a potential crisis of over-accumulation in capitalist economies. Over-accumulation in the primary sector (which includes industry) was resolved, it is argued, by capital 'switching', through the economic policies of central governments and the actions of lending banks, to the secondary sector, which includes property development. This, they argue, created a boom in property development, fuelling suburban expansion.

By contrast both humanist and feminist geographers have developed critiques of the landscapes and cultural assumptions which have underpinned suburban development. Humanists have highlighted the lack of originality in the design of suburban landscapes, their monotony and lack of regional distinctiveness. This, they have argued, has failed to provide these landscapes with any 'sense of place', a vital psychological component of human feelings of security and belonging. This perspective has been criticised by sociologists and geographers, who argue that this is an elitist perspective which ignores a variety of attachments to ordinary landscapes that belies their uniform physical appearance.

Feminist geographers have attacked the sexist stereotyping inherent in the production and promotion of suburban landscapes. The design and depiction of the suburban house and its landscape promote notions of domesticity, consumption and recreation. This, they argue, hides the fact that these landscapes are largely maintained by women, who, because of male dominance of car use and poor public transport provision, are daily trapped in these areas. The image of suburbia which hides the work required to maintain the ideal of this landscape is part of the general failure to regard housework as 'real' work. The feminist critique provides a valuable corrective to idealised representations of suburban life.

Post-suburban developments in the UK

Observers of the landscapes around the edges of major cities in the USA have begun to recognise the emergence of a new form of residential landscape that has been termed 'post-suburban'. These developments include extensive, private, master-planned developments that are radically different from the traditional residential suburb. They are aimed at the affluent consumer. Individual dwellings are typically large and situated in extensive, well-landscaped grounds. The design of individual dwellings and the landscaping of whole developments are closely based on ideas of tradition and rurality borrowed heavily from idealised anglophile notions of the rural realm (Knox 1992a). These developments are marketed and named accordingly, in formerly undeveloped plots, and come provided with a wide range of amenities, normally found in towns, which are planned into them. These include a town centre, public squares, police and fire stations, libraries, theatres, meeting halls, schools and post offices. All of these are designed to fit into the overall motif of the development. They surround, or are found in close proximity to, significant concentrations of offices, shops and other formerly 'downtown' commercial functions which have decentralised, driven by high land costs and central city disorder, and freed up by telecommunications developments. These, rather than being an extension of the urban area, which is the case with conventional suburbs, may be regarded as an alternative to the city; they have come to be known as 'edge-cities' or 'stealth-cities', a reference to the fact that they frequently straddle a number of administrative districts and, therefore, fail to crop up on official statistics and census returns (Garreau 1991; Knox 1992a, 1992b, 1993).

What is distinctly different about these developments is their origins in the private sector, their development which was fuelled entirely by the development boom of the 1980s and the control exerted over their design by a single developer. In the USA they have rapidly become significant elements of the urban geography of North American cities. The proliferation of these developments around major cities has somewhat dissolved their traditional form and created an urban form which has been referred to as a 'galactic metropolis' (Lewis 1983; Knox 1993).

The immediate fringes of British cities and the areas beyond have been the locations of considerable development, with, for example, the decentralisation of retail and industrial land uses, the development of business and science parks and the continued pressure of the outward spread of housing. The most developed example of what should be called the 'spread-city' in the UK, rather than the edge-city, is London and the surrounding South East region. A considerable amount of the economic activity of the South East region has developed along axes from London. These have included the three motorway axes: the M3 axis from London to Southampton which includes Guildford, Basingstoke and Southampton, the M4 axis from Heathrow Airport through Slough and Reading towards Swindon and extending to Cheltenham and Bristol (this area contains around 60 per cent of the new high-technology firms established since the mid-1970s), and the research and development and aerospace-oriented axis along the M11 through Hertford towards the Cambridge Science Park established by Trinity College. The restructuring of these areas is closely linked to the geography of government research centres which include Aldermaston (weapons research), Farnborough (aircraft) and Harwell (nuclear). These areas, as well as a number of historic towns in the North, such as York and Lancaster, have developed economies based around technologically advanced methods of production and social profiles showing high numbers of the emergent, affluent 'service class' (employers, managers and professionals). Such developments have had a significant impact upon the British urban scene. However, they have contributed to a general decentralisation of British urban form, rather than the dissolution that geographers have argued is the case in the USA. Their regional concentration in the South East and limited impact upon the overall form of other British cities appear to mark these developments out as an exception within British cities rather than the harbinger of the age of the British edge-city. The rediscovery of the city centre as a focus for urban cultural life and the predications that the actual amount of telecommuting in the future is never likely to match the potential which technology offers would support the notion that the traditional focus of British urban form has not yet been superseded by a post-suburban equivalent. The differences between the urban systems of the USA and Europe suggest that, while the edges of European cities are increasingly importanct foci for the study of urban development, simply importing theoretical perspectives derived from the study of US post-suburban development is unlikely to prove enlightening. The differences between Europe and North America and the differences within European urban systems need to be appreciated to understand new urban forms appearing around the edges of existing cities (Keil 1994).

Postmodern urbanisation: a blueprint for urban change?

This chapter has been concerned with an apparent transformation ('postmodernisation') of the processes of urbanisation and some of the resultant forms. The products of this postmodernisation are very apparent in parts of North America, especially in the hi-tech-oriented world cities such as Los Angeles and Washington, DC. Even a cursory review of the British landscape, for example, reveals that such postmodern urban landscapes have not yet emerged on such a large scale. London is the best example of a city whose centre has been transformed along the lines seen in other world cities around the world. This is closely related to its position as a, possibly *the*, pre-eminent world financial centre, while its suburbs and surrounding regions are the best exemplars of the economic and social restructuring associated with the rise of hi-tech industry. This is related to its wealth of research facilities, the presence of an appropriate workforce and the pleasant suburban environments in which they live. However, these elements point to the fact that these areas are largely the exception rather than the norm within the British urban system. Certainly, examples of postmodern urban form may be found elsewhere: the service-oriented yet 'historic' settlements of Lancaster and York, traces of postmodernism within otherwise largely 'modern' city forms, and the spectacular redevelopments of city centres surrounded by areas of social and economic deprivation. However, it would be inaccurate to declare the British urban landscape unequivocally postmodern; whole areas of many British urban settlements have been largely unaffected by the apparently epochal forces of postmodernism. Likewise, the partial, rather than total, transformation of urban form is the case for most cities in Europe and Australia. Postmodern urbanisation, like most other facets of urbanisation, is emerging as a complex series of trajectories mediated locally rather than a single, simple, universal trajectory of development.

There are three reasons for this state of partial transformation (or rather one reason expressed three different ways). First, the processes of postmodern urbanisation are general processes; however, they are mediated by local conditions. Their outcomes depend upon the mediating effects of local social, economic and cultural conditions, and the effects of individual actors in particular situations. The existence of general processes of postmodern urbanisation does not automatically and unproblematically translate into the production of postmodern cities. Second, cities are not passive receptors of change. While general processes of urbanisation may try to make their imprint on urban landscapes they meet with resistance and are tied by the legacy of existing urban landscapes. Cities are not easily and infinitely flexible and they exert a considerable influence over the operation of general processes of urbanisation. Again the operation of general processes of postmodern urbanisation does not necessarily translate into urban transformation. Finally, urban hierarchies and individual cities are internally heterogeneous. General processes of urbanisation are refracted through these patterns of internal differentiation. Consequently, while some cities and some areas of cities are affected by the processes of postmodern urbanisation, others are not. In conclusion then, the notion that we are witnessing a general urban transformation needs to be qualified. The new processes of urbanisation that have been recognised as the causes of this transformation actually produce a geographically uneven pattern of 'transformation' which varies regionally, between cities and within cities.

Project idea

Collect a range of evidence from a city you know well to evaluate the view that postmodern urbanisation is dissolving the structure of cities. This evidence may include the structure of the town or city and the location of key activities and the nature of major new developments. You might use evidence from the census over a period of years to map levels of social and economic welfare. What evidence can you find to suggest that cities are being transformed by postmodern urbanisation?

Essay titles

- To what extent, and in what ways, has fear and paranoia become manifest in the urban landscape?
- Mythological views of suburbia hide a complex and contested reality. Discuss.

Further reading

Byrne, D. (2001) *Understanding the Urban*, Basingstoke: Palgrave (chapters 5 and 7). [Concise discussions of a number of issues of relevance to this chapter from a sociology perspective.]

Eade, J. and Mele, C. (eds) (2002) *Understanding the City: Contemporary and Future Perspectives*, Oxford: Blackwell. [Wide-ranging, international collection of essays blendng theory with empirical analysis.]

Knox, P. (ed.) (1993) *The Restless Urban Landscape*, Englewood Cliffs, NJ: Prentice-Hall. [Although a little dated now, still an excellent and very relevant collection of essays.]

Knox, P. and Pinch, S. (2000) *Urban Social Geography: An Introduction*, Harlow: Prentic-Hall (4th edn). [A number of key discussions of the social dimensions of urban change.]

Silverstone, R. (ed.) (1997) *Visions of Suburbia*, London: Routledge. [A diverse collection of international essays on cultural aspects of suburbia.]

Web resources

European Spatial Planning Research and Information Database – www.esprid.org

Re Urban Mobil – www.re-urban.com

Urban Institute – www.urban.org

8 Unequal cities

Five key ideas

- **Regeneration projects have had positive impacts upon the landscapes of cities, levels of inward investment and civic pride.**
- **The apparent economic regeneration of many cities may be challenged on a number of fronts.**
- **Levels of socio-economic division in many cities have worsened despite being subject to extensive regeneration.**
- **The regeneration of many cities has been subject to only limited local democratic accountability.**
- **The fashioning of new cities through regeneration has been subject to resistance and protest.**

Introduction

It should be apparent that cities have been forced to adjust to severe structural changes occurring within national and international economies. Some of the most heavily affected are the former manufacturing cities of various rust-belts, in the north of the UK, north-east USA and certain Australian states. These cities have been at the sharp end of the effects of the globalisation of the world economy and its transition from industrial to post-industrial. They have seen their economic bases largely eroded and with them their former affluence and economic pre-eminence. How they have adjusted to these wider economic changes is of profound importance to their current and future welfare and development. This chapter offers an evaluation of the measures undertaken by these cities to adjust to changing economic circumstances.

Evaluating the regeneration of the city

Previous chapters have outlined various ways in which the management of cities has broadly become more entrepreneurial and enterprise oriented. The focus of urban government has shifted away from managerial, welfare issues towards those of promoting economic development through regeneration. This tendency has been particularly well developed in many British, European and North American rust-belt cities. 'Regeneration' should be regarded as a problematic term and consequently viewed critically. The question 'What does regeneration mean?' is an important one, yet it is ill considered by many politicians and policy-makers. All too often regeneration is equated with economic or physical regeneration. This in turn is equated with an increase in wealth or jobs in an area. However, this is an inadequate, or at the very least narrow, definition of regeneration. This definition ignores a number of issues which are crucial to the evaluation of the success or failure of projects of urban regeneration.

The first issue concerns the characteristics of this wealth or these jobs. For example, what is the distribution of wealth among different social groups? Is it equal or fair? What are the type and quality of jobs stemming from urban regeneration? What is the distribution of these jobs between different social groups? The second issue concerns whether or not all of the impacts of urban regeneration are positive. Are there any negative impacts of urban regeneration? If so, how are these negative impacts distributed between different social groups?

The aim of these questions is to open up the claims of urban regeneration and its rhetoric to critical analysis. This chapter evaluates the claims of urban regeneration and assesses its economic, social, political and cultural impacts upon cities and their populations.

Economic impacts

At first glance spectacular projects of urban regeneration appear to offer attractive alternatives to the derelict industrial land that many are built upon. As conspicuous displays of wealth and with their emphasis on high-quality urban design, they create the impression of revival. They are confident architectural statements. They also offer a number of direct and indirect benefits to the urban economy. By being 'people-attractors' of some description these projects lure a significant number of visitors to

Case study G

Glasgow: new image, old problems

Glasgow has undergone extensive regeneration through programmes including the 1988 Garden Festival, the 1990 European City of Culture and as the UK City of Architecture and Design in 1999. In addition, it is now generally perceived by outsiders as a modern, stylish and cultured city. In March 2004 the latest chapter in the re-invention of Glasgow was the launch of a new image under the strapline 'Glasgow: Scotland with Style'. However, despite these efforts, basic levels of health and economic and social welfare in Glasgow remain worrying low or are worsening for many of its poorest populations. For example, life expectancy among men in some of Glasgow's poorest wards is only 63, fourteen years below the national average. Much of the city's public housing is in poor repair and of Glasgow's 90,000 unemployed, three-quarters receive sickness benefit. While the economy, particularly in the central area, continues to boom, driven largely by tourist revenues, there are significant proportions of the city's population who seem permanently cut off from the benefits of this economic renaissance. Critics argue that rebranding alone, while it may bring economic rewards, is insufficent to address the most intrackable problems afflicting the poorest urban populations. For example, Tommy Sheridan, leader of the Scottish Socialist Party, has argued: 'We are a very vibrant city, there is no doubt about it. I love Glasgow. It's my city. But we have huge problems and I don't think the rebranding will tackle those problems.'

Source: Scott (2004)

cities. Bradford's National Museum of Film and Television is a good example. This development has been instrumental in virtually creating a tourist economy in Bradford out of nothing (Bianchini *et al.* 1992: 251). Recorded visits to Bradford increased from none in 1980 to well over five million ten years later. The beneficial spin-offs from this cannot be underestimated. Spending in the city, and others such as Glasgow which have adopted similar approaches to regeneration, has increased significantly in retailing, entertainment, catering and hospitality, hotels and other consumer service industries. As a result the benefits of regeneration have spread throughout the city centre, transforming the appearance of many older and formerly derelict buildings. Old buildings have been renovated and refurbished, and new ones built to tap into the expansion of the tourist economy. Moreover, one of the prime functions of these types of development has been to raise the national and international profile of towns such as Bradford or at least ensure that they are known for positive rather than negative reasons (Bianchini *et al.* 1992: 249–51). City leaders and promoters refer to this as 'putting the city on the map'. This has been important in attracting the attention of national and international organisations such as national tourist boards and the European Union, particularly where grants or financial help have been available to subsidise these developments. However, benefits such as these should be balanced against the complex, multifaceted and interrelated nature of problems in the city. The example of Glasgow demonstrates the limits of regeneration programmes. An evaluation of the impacts of urban regeneration should ask: How and why does regeneration fail? And who does regeneration fail?

To recap: the economic rationale behind regeneration is for a redistribution of income within the city via a combination of 'trickle-down' and the multiplier effect. The assumption – for there is little actual evidence of its effectiveness – is that increased revenue from visitor spending and investment will do two things. First, it will trickle down into the pockets of the most disadvantaged through the creation of jobs servicing visitors and incoming investors (Hambleton 1995). Second, it is assumed that this revenue will have a positive, knock-on effect as it spreads through the local economy as a result of increased consumer spending. These assumed economic mechanisms are held up as justification of the claims of urban regeneration.

However, this rhetoric, and its inherent assumptions, ignore a number of critical issues. Perhaps most importantly they ignore the compelling evidence against the effectiveness of the trickle-down and multiplier mechanisms. Further, they ignore the wide-ranging negative impacts that projects of urban regeneration can have on the local economy and upon certain groups within it. Clearly in any genuine evaluation of the impacts and effectiveness of urban regeneration, rational debate needs to replace the rhetoric that has been characteristic.

As mechanisms of income redistribution advantaging the poorer sectors of urban societies, both the trickle-down and multiplier effects are likely to be very ineffective. The evidence suggests that either the increased revenue in urban economies as a result of urban regeneration will fail to 'drain' out of the business and managerial sectors, or it will do so through those channels that fail to benefit poorer populations substantially. These channels include the creation of low-paid, unskilled, insecure, part-time employment. These are likely to be the only employment opportunities stemming from projects of urban regeneration that are open to poorer populations (Loftman and Nevin 1994). I will return to the economic effects of these types of jobs later in this chapter.

Likewise, any multiplier effect stemming from urban regeneration is likely to be very low. This is because the majority of goods purchased as a result of increased consumer spending are unlikely to be manufactured locally. The result is that money 'leaks' out of local economies. If increased consumer spending is concentrated on imported goods, this may lead to a worsening of balance of payments nationally (Turok 1992).

The apparent benefits of urban regeneration also vary with the scale at which the issue is examined. Projects of urban regeneration tend to be spatially autonomous; they are designed to regenerate specific spatially defined areas. The result of the proliferation of a large number of very similar projects of urban regeneration is that, when examined on a national scale, they are likely to be in competition with each other. While growth may occur in one area, it is likely that this will be at the expense of decline in another area. When examined on a national scale, projects of urban regeneration may lead to no overall growth. If they have any national economic effect it is likely to be simply the creation of a new pattern of growth areas and areas in decline. This phenomenon is referred to as 'zero-sum growth' (Harvey 1989b).

Despite the evidence, the rhetoric of regeneration tended to hold sway over rational planning and debate for much of the 1980s and early 1990s. Projects of urban regeneration are rarely accompanied by impact assessments, and specific, effective mechanisms of income redistribution are rarely studied or put into place.

The predominant model of urban regeneration in North America and Europe in the 1980s and 1990s was centred upon some form of property development or regeneration. This is based upon the belief that a strong link exists between property development and economic regeneration. However, a fundamental question mark hangs over the effectiveness of this link. Property is a sector of the economy that has never displayed any degree of long-term stability. It is characteristically susceptible to economic swings and fluctuations in property values. Investment in land and property is inherently highly speculative and the results are far from guaranteed (Turok 1992; Imrie and Thomas 1993).

Employment issues

Effective and sustained recovery requires that localities do more than just create the impression of recovery. Not only must the surface appearance of an area be transformed but so must the underlying economic characteristics that created the need for regeneration in the first place. The impacts of national or international economic decline are mediated at a local level. Their outcomes are not fixed or predetermined. Rather they depend upon the interaction of wider processes and local characteristics, be they economic, social, cultural and/or political. If these local characteristics are left untouched by regeneration, then the future fortunes of places in relation to the impact of external forces are likely to remain similarly untouched. The conclusion must be that regeneration of this type is largely cosmetic.

Major problems which determine the local outcomes of external impacts include a lack of appropriate skills, training and educational qualifications combined with the lack of a sound local economy characterised by secure, skilled, well-paid, long-term employment opportunities, investment and innovation among local companies. To assess the degree to which the 'substance' of regeneration matches the appearance, it is important to consider the issues of the quality of employment, training, opportunity and investment present within local economies. How, to what extent, and in what ways are the local economy and labour force strengthened by regeneration?

One of the major claims of property-led models of urban regeneration is that they create jobs. These jobs stem directly from the construction of properties and indirectly in the running of developments once constructed. However, there are a number of characteristics of the construction industry that limit these benefits. Few construction jobs tend to be recruited from local populations; normally construction workers migrate to jobs from elsewhere (Turok 1992: 362–3). Even in cases where legislation has been introduced to try and encourage the local recruitment of construction labour it has tended to be unsuccessful, falling foul of official guidelines on discrimination (Loftman 1990). In addition, the nature of employment in the construction sector is such that jobs tend to be irregular and short term. The construction sector traditionally has a very poor record in training. Employment in the construction of property linked to urban regeneration, while it may alleviate short-term hardship, is unlikely to do anything to equip the population with useful or marketable long-term skills. Furthermore, job opportunities in the construction sector tend to be restricted to young males; therefore, large sections of the population in need of the benefits of regeneration are excluded (Turok 1992).

Serious doubt has also been cast over the importance of property in the relocation decisions of companies. Social rather than physical factors tend to be important in these decisions. Factors such as perceived quality of life appear more important considerations in the long-distance relocation decisions of companies (Duffy 1990; Turok 1992). Consequently, developing a supply of office space may not be a sufficient lure for firms to relocate from elsewhere. What is more likely to occur is firms moving shorter distances within labour markets, another form of zero-sum growth. Where relocations from outside local labour markets to newly developed properties have occurred it has tended to be predominantly lower-order functions (clerical and administrative) rather than higher-order functions (management, control and ownership) (Turok 1992; Graham and Marvin 1996). These higher-order functions tend to display a greater inertia than do lower-order functions. The result is the danger of a form of 'branch-plant' urban economy developing. This type of economy provides little indigenous momentum for future growth and causes urban economies to be vulnerable to decisions made elsewhere.

Problems have also stemmed from a lack of originality in the urban regeneration schemes undertaken by cities in the USA and Europe. They have typically consisted of the serial reproduction of a few models of urban regeneration. These models have usually consisted of some combination of hotel developments, exhibition or convention centres, retail parks, heritage sites, waterside developments, offices and luxury residential developments. The market for these developments is not infinite. Such unimaginative reproduction of physical developments runs the risk of creating market saturation and a consequent underuse of capacity. Such a possibility is obviously in conflict with the aims of those promoting regeneration (Bianchini *et al.* 1992: 254).

> 'Heightened inter-urban competition produces socially wasteful investments that contribute to rather than ameliorate the over-accumulation problem. . . . Put simply, how many successful convention centres, sports stadia, disney-worlds, and harbor-places can there be? Success is often short-lived or rendered moot by competing or alternative innovations arising elsewhere.'
>
> (Harvey 1989b: 273)

A further problem stemming from the limitation of standard models of regeneration is that these developments are rarely well tuned to the specific nature of the localities within which they are located. The potential mismatch may express itself in a number of ways. Most obviously it presents a mismatch between the opportunities available for local people, their employment and training needs, and their present skills. This failure to use local expertise is socially wasteful and an inefficient use of human resources.

Undoubtedly, projects of urban regeneration generate jobs within developments themselves. However, two qualifications should be applied to this statement. First, developments might have a negative impact, either locally or across wider areas, causing jobs to be lost elsewhere. Second, there is a question mark over the quality of jobs created in these developments, and their ability to alleviate conditions of poverty locally.

Projects of urban regeneration once opened up may be in competition with other facilities locally. The result of this is that their success may be at the expense of job losses elsewhere in the local area. This has been most apparent where retail developments have been used as the basis of urban regeneration. The impacts of massive regional shopping developments – such as Merry Hill, near Dudley, in the West Midlands, and the Metro Centre in Gateshead, near Newcastle upon Tyne – on nearby older retail environments has been severe. Rather than bringing increased wealth and prosperity to regions, they may initiate new patterns of uneven economic development as well.

The patterns of employment created in urban redevelopments tend to be highly polarised. They have been characterised by a relatively small number of highly paid managerial jobs and a larger number of much less-well-paid, unskilled jobs in sectors such as catering, security and cleaning. This pattern of employment has led to what has been referred to as a bifurcation of opportunity (Short 1989). There exists a clear qualitative and quantitative mismatch between the few highly paid managerial jobs available in urban regeneration schemes and the needs and skills of local poorer populations. Consequently these jobs tend to go to well-qualified outsiders. Local populations tend to find their opportunities restricted to the less-well-paid, insecure sectors of the urban economy.

Bearing this bifurcation of opportunity in mind, it is apparent that there are serious doubts over the quality of opportunity available to poorer urban populations, those, in theory at least, with the greatest need. The jobs available to disadvantaged residents are typically unskilled, low paid, temporary or short term, non-unionised and offering poor-quality training (Loftman 1990).

Local authorities have not been blind to these criticisms. They have often tried to overcome them by introducing legislation that targets the disadvantaged populations and links jobs to training programmes. However, in practice these measures have yielded disappointing results. The number of people who have benefited from them has been very low on the whole. Despite local authorities' good intentions this legislation has largely failed to overcome the limitations of the employment opportunities linked to projects of urban regeneration.

Displacement issues

Urban regeneration often involves the redevelopment of extensive areas of land. This has a considerable impact upon the existing occupiers of these and other sites nearby, be they industrial, commercial or residential. There are many documented cases of the severe impact of urban regeneration on existing businesses. This is especially the case where these businesses have been regarded as incompatible to the image being imposed upon an area by regeneration. This impact is most often manifest in displacement pressures. Such pressures may stem from clean-up programmes designed to enhance the image and appearance of redevelopment areas. Displacement may have a number of adverse effects on the businesses affected. These effects include the breakup of local business networks and connections with other businesses, customers, suppliers and markets, a lack of appropriate property available elsewhere and an overall poor location. These impacts have been severe enough in many cases to lead to the closure of displaced businesses (Imrie *et al.* 1995: 34–5).

Urban regeneration also has a displacement potential for residential populations. This is examined in more detail under the heading 'Housing issues' (below).

Subsidy issues

Urban regeneration in British cities in the 1980s and 1990s was characterised by partnerships between public and private sectors. In effect this means that private sector development has been heavily subsidised by public money. This subsidy may involve a direct financial grant to private developers, for example local authorities undertaking some of the construction costs of developing a major international hotel, local authorities developing facilities, such as convention centres, which are likely to be used primarily by the private sector, or relaxing local rates or taxation to encourage business relocation. This public sector subsidy has come at the same time as a severe restriction of money to local authorities both through central government grant and money raised by local authorities through local taxation (Goodwin 1992). The subsidy of private sector development, therefore, may impose both a severe short-term cost on local authorities and a more long-term one as they pay back loans over periods of up to twenty-five years (Loftman 1990). This has resulted in some cases in local authorities raising taxes from local residents to subsidise a relaxation of costs on the private sector. This diversion of public funds has had far-reaching implications for the management of urban areas.

The best documented problem stemming from the diversion of public funds has been the effect this has had on social spending by local authorities. Sectors such as housing and education have often been characterised by under-funding in large urban areas in the 1980s and 1990s. This was particularly severe in disadvantaged inner-city locations. These costs have tended to be borne by the more disadvantaged populations who typically have a higher dependency on the public sector provision of housing, health and education.

The nature of the types of urban regeneration projects that these moneys have been invested in are inherently speculative. There is no guarantee, therefore, of their success, or that the public money spent will be returned (see e.g. Wynn Davies 1992). Some ventures, for example convention centres, are not designed to make money themselves. They act as 'loss-leaders' encouraging increased expenditure by visitors in other sectors of the urban economy. In effect the public–private partnership that has characterised urban regeneration in the UK involves the public subsidy of private profit (Harvey 1989b). Further to this, a number of critics have questioned the need actually to subsidise private investment. They argue that in many cases private investment does not necessarily respond to public subsidy (Turok 1992: 374–6).

Problems with the type of investment encouraged

The type of investment targeted by urban regeneration projects is inherently unstable. It is likely to fluctuate with swings in the economy. This is the case, for example, with the business-tourist market. Such investment is also geographically footloose and is prone to rapid switches of location in response to minute variations in the social characteristics of locations.

The retail sector has formed an important part of many urban regeneration developments. These have included regional shopping 'mega-malls', such as the Metro Centre in Gateshead, and festival shopping developments, such as those in the converted warehouses of London's Docklands. However, certain characteristics of the retail sector severely limit its effectiveness as a promoter of sustainable, fair economic regeneration. Employment opportunities in the retail sector open to disadvantaged populations are limited. They are typically of a low quality, fitting the profile of poor employment opportunities outlined earlier in this chapter. Encouraging retail developments also engenders potential problems for consumers. The retail boom of the late 1980s was fuelled largely through a boom in the availability of personal credit. This led later to many problems of personal debt as consumers struggled to repay the amounts they had borrowed plus the associated interest (Turok 1992).

Summary: economic issues

First, urban regeneration projects engender a highly uneven distribution of costs and benefits. The benefits tend to accrue to a small number of professionals from the business and managerial classes. These are often the people who would have done well anyway, without regeneration. The groups in most need of economic benefit, the disadvantaged populations occupying de-industrialised inner-city areas or run-down peripheral housing estates, tend instead to bear the brunt of the costs and negative impacts of urban regeneration.

Case study H

Urban renaissance: myth or reality in Cleveland, Ohio?

During the 1960s, 1970s and much of the 1980s, Cleveland, Ohio was regarded as the epitome of all that was wrong with urban America. It was polluted, so much so that its river actually caught fire in June 1969, its economy was bankrupt, its leaders were widely regarded as incompetent and provincial, and its baseball team had not graced the World Series for over thirty years. The city's epithet said it all: 'The mistake by the lake.' However, by the 1990s the talk of Cleveland was of a city reborn. The most tangible symbol of the alleged rebirth was its landscape. This included a $200 million stadium for the now successful baseball team, a postmodern showcase development housing the Rock and Roll Hall of Fame, as well as a host of other impressive architectural developments. What is more, herons can land on the Cuyahoga River without fear of getting their feathers singed.

Cleveland's apparent transformation dates back to the conversion of the terminal of the Baltimore and Ohio Railway into a leisure, retail and commercial development in the early 1980s. The process has since involved the business community, non-profit foundations and new municipal leaders.

Cleveland's renaissance has not passed without criticism, however. Clearly the renewal projects have been undertaken with particular audiences in mind. These include middle-class suburban Cleveland rather than its poorer inner city. For example, while the city's downtown has been extensively revamped, the same could not be said of its school system (Cornwell 1995).

While Cleveland looks like a city on the up, the extent to which all in Cleveland could be said to be sharing this trajectory is more questionable. Cleveland is an example of an American 'urban success story', a city trumpeted by the media, marketers and politicians as a successfully revitalised city. A number of other cities in the USA and Europe have made similar claims. However, such claims are typically light on objective, or tangible measures of regeneration. The question that remains is: How real is this regeneration? The question of what is meant by real is clearly problematic and unlikely to draw universal agreement. However, a basic requirement of a claim to meaningful, 'real' regeneration might be that the economic well-being of residents has improved. While this is clearly not the only measure that could be employed, it appears to be a fundamental one.

To evaluate the claims of these urban success stories for North American cities it was necessary to draw up a list of cities that were 'distressed' in 1980. This was done using an index that included measures of unemployment, poverty, household income and income change, and population change. From this a list of fifty economically 'distressed' cities was compiled. It was necessary next to find out which of these cities had been held up and widely regarded as a success story since 1980. To discover this a range of expert opinion was polled; these experts included academics and economic development practitioners. From their responses two classes of 'successful' cities were drawn up: twelve 'successfully revitalised' cities, mentioned by 20 per cent or more of respondents and of these six 'most successfully revitalised' cities, mentioned by 40 per cent of more of respondents. The 'most successfully revitalised cities' were Pittsburgh, Baltimore, Atlanta, Cleveland, Cincinnati and Louisville. This was based on the respondents' perceptions rather than any objective measures of revitalisation. It was important to assess the extent to which these perceptions accorded with more objective measures of revitalisation.

To evaluate the claims of these success stories the 'successfully revitalised' cities were compared to those cities deemed 'unsuccessful'. An index of the performance of these cities between 1980 and 1990 was constructed using measures of residents' economic well-being. The results of this evaluation were revealing. They showed that there was no significant difference in the performance of 'successfully revitalised' cities compared to 'unsuccessful' cities. Stories of urban revitalisation and renaissance appeared not to translate into tangible improvements in the economic well-being of the residents of these cities, compared to those of 'unsuccessful' cities. In fact many 'unsuccessful' cities performed better than 'successful' ones over the period on a majority of indicators. Of the six 'most successfully revitalised' cities, only Atlanta and Baltimore significantly outperformed the 'unsuccessful' cities; of the 'successfully revitalised' cities, only Boston achieved this.

Cleveland's performance in this evaluation paints a rather different picture from that of the press report discussed above. Cleveland performed significantly more poorly than the average of 'unsuccessful' cities on all indicators used: percentage point change in unemployment, percentage change in median household income, percentage change in those living below the poverty line, percentage change in per capita income, and percentage point change in labour participation rate.

Two conclusions may be drawn from this evaluation of 'urban success stories'. First, such success stories appear to be based largely on image and perception rather than on objective measures of well-being. Second, those cities heralded as urban success stories tend to be those which have displayed an 'appearance' of revitalisation, namely those cities which have physically upgraded their central areas and enhanced their images. However, there is no necessary correspondence between an appearance of revitalisation and its realisation in terms of residents' economic well-being. Cleveland's claims to regeneration appear particularly hollow in this light.

Sources: Cornwell (1995) and Wolman *et al.* (1994)

Social issues

The dual city?

The bifurcation of economic opportunity described in the previous section has had an inevitable influence upon the social geography of cities (Short 1989). A metaphor used frequently to describe these new social geographies is that of the 'dual city'. The idea of a dual city is based upon evidence of increasing social division within cities and the apparent emergence of an urban 'underclass' divorced from dynamic mechanisms in the formal economy. This economic exclusion is translated into exclusion from many areas of city life. This underclass consists of both waged and unwaged poor, a disproportionally high number of members of ethnic minorities, and groups such as sick, elderly or disabled people and single parents. It has been argued that social polarisation has increased as a result of both international economic trends and recent government policy. This social polarisation has had an impact upon the spatial structure of cities. Areas such as inner-city areas and peripheral council housing estates have become equated with the presence of this urban underclass. These pockets of deprivation contain a number of groups who are surplus to the requirements of the formal urban economy. These areas are

Plate 8.1 ***Run-down housing adjacent to new hotel developments, Cultural Forum 2004 site, Barcelona***

characterised by the prevalence of 'alternative' or 'twilight' economies run on an informal or illegal basis.

These pockets of deprivation have become contrasted to highly spectacular 'islands' of regeneration which exist often in very close proximity to these 'seas of despair' (Hudson 1989). London's Docklands provides an example of just such a spectacular island of regeneration set among some of the poorest communities in London. In Birmingham (UK) the International Convention Centre built at a cost of £180 million is only 200 yards from Ladywood, one of the city's poorest wards (Loftman and Nevin 1992). While the picture painted here is something of a generalisation and the unique characteristics of individual cities should not be ignored, it does highlight an important trend in the social geographies of urban areas in Britain and to a greater degree in the USA.

Urban regeneration and social policy

Social improvement appears to be the indirect, rather than the direct, consequence of urban regeneration projects. The social and community policy aspects of urban regeneration programmes have typically received a very low proportion of total regeneration spending . They also tend to be the most vulnerable aspects of these programmes during times of economic recession.

There is little evidence that those outside the formal job market have been included in the social regeneration policy aspects of urban regeneration programmes. Social regeneration has been all too frequently equated with physical and economic regeneration.

Housing issues

Urban regeneration projects affect the local housing stock in two ways. These are physical displacement of poorer resident populations and displacement through forcing house price increases. Physical displacement may occur where residential properties are demolished to make way for newer developments. It is, inevitably, cheaper, low-value accommodation that is removed and which serves low-income, marginalised populations. In some cases hotel and rooming accommodation which served as a valuable source of cheap accommodation has been renovated and modernised to cater for new, often tourist, markets which are expected to flood into areas following regeneration. The consequences of this have been that poorer residents have been evicted in the short term and have suffered a drastic reduction in the supply of affordable accommodation available to them in the longer term. The physical and psychological impacts of this have been severe on residents who may have been elderly or suffering from long-term illnesses. Frequently there has been no council housing provision to compensate for displacement. Displacements such as these have been common in the large-scale redevelopment of North American cities, for example in association with the development of Vancouver's Expo '86 site (Beazley *et al*. 1995).

Low-income populations may also suffer displacement through being squeezed out of local housing markets as regeneration forces the value and hence the price of local housing up and out of the range of low-income populations. Areas around regeneration projects have frequently been subject to an increase in property speculation and interest from property developers with an eye on encouraging the influx of middle-class professionals with a taste for city centre living. This interest adversely affects the ability of low-income populations to purchase property locally, forcing them into alternative cheaper property markets, often in different localities. This is an example of a process known as gentrification. Although by no means caused exclusively by urban regeneration, it may be exacerbated by it. This has proved a greater problem than direct displacement in British cities to date. Other development pressures suffered by those adjacent to redevelopment projects have included the loss of community facilities to make way for redevelopment and disruption and noise pollution during the construction phase of development.

Flagship projects of urban regeneration frequently include programmes of housing provision or renovation. Waterfront developments typically use the legacy of distinctive architecture and original features, converting them into, for example, luxury housing. Again, they are aimed at luring incoming professionals. This accommodation does nothing to solve housing and accommodation crises among local poor populations, who are unable to take advantage of these developments. Housing developed in association with such projects of urban regeneration tends to be inappropriate to local populations on a number of counts. First, it is far too expensive. The majority, often over 80 per cent, of this housing is owner-occupied. During the housing boom of the late 1980s some two-bedroom flats in London's Docklands were selling for £200,000, while the majority of households in the surrounding boroughs had combined annual incomes of less than £10,000. Further, the types of properties provided in these developments are aimed primarily at the 'dinky' market (double income, no 'kids', typically young professional couples in their twenties and early thirties). Consequently this accommodation consisted primarily of small flats. The accommodation needs of the majority of local households are usually very different from those of the 'dinky' couple. Often families are older and with children. They require housing with three or more bedrooms and preferably gardens and play areas. The areas surrounding inner-city regeneration schemes in the UK often contain a large proportion of families from Asian ethnic minorities. These communities are characterised by a high proportion of extended family households. Their requirements are again for the larger housing units notoriously absent from urban regeneration schemes.

The provision of 'social' housing, affordable accommodation aimed at addressing local housing needs, has long been a major plank of the social regeneration programmes of urban regeneration schemes. In Cardiff Bay, for example, it was agreed that 25 per cent of new housing in the development should be social housing. However, the actual delivery of this social housing has tended to be very disappointing. Frequently, there is little attempt to match the housing provided with the characteristics of the local population. All too often, as well, the actual amount of social housing in completed regeneration schemes is well below the amount initially agreed (Rowley 1994). Often the need to generate profit from housing ultimately overrides social considerations.

Discussion topic

The regeneration of city centres has done little if anything to improve levels of social and economic welfare for the most needy urban populations. Is this an accurate reflection of the impacts of recent rounds of urban regeneration? What evidence do you have to support your answer?

Issues of local democracy and public involvement

Urban regeneration has taken place within changing structures of urban governance. The creation of agencies, such as urban development corporations and training and enterprise councils in the UK, has effectively taken certain areas out of the control of local authorities and transferred accountability away from the local area and towards central government. This and the other political characteristics of urban regeneration have been accused of imposing a barrier to public involvement in decision-making, and ultimately a threat to local democracy as it has developed in British cities.

Projects of urban regeneration in the UK, Europe and North America have been characterised by a lack of public debate, consultation, inquiry or detailed prior impact assessment. The exclusion of the public from the development process in this way has been facilitated by wider political and legislative frameworks which are biased in favour of 'growth-coalitions' and against the inclusion of public and opposition groups (Beazley *et al.* 1995). Debates, for example on the costs and impacts of urban regeneration, which, prior to the introduction of centrally appointed agencies such as urban development corporations, were conducted in the public sphere, now take place as part of private board meetings (Imrie *et al.* 1995). Furthermore, legislation in the Local Government Act 1985 allowed information to remain confidential and away from public scrutiny (Beazley *et al.* 1995).

In the face of powerful regeneration coalitions supported by legislation, community groups and opponents to regeneration appear weak, under-resourced and distanced from power. In financial terms they are unable to match the powerful regeneration coalitions: in the majority of cases these groups are funded by donation or subscription. Occasionally they are funded by grants from private companies or local authorities. However, this has been a source of controversy as these are often the very organisations that are being opposed. Opposition groups are further excluded from debate by the restricted and limited information divulged by regeneration coalitions, a lack of legal or technical support afforded to opposition groups in comparison to that available to regeneration coalitions and the increasingly 'closed' nature of decision-making in local government. When faced by apparently unified regeneration coalitions, opposition groups appear fragmented because of their diversity of agendas. Consequently public involvement in regeneration programmes has been extremely limited. While local authorities and regeneration coalitions might promote the appearance of public involvement through, for example, public meetings, these are often well after development has begun and they have no influence on decision-making or policy (Beazley *et al.* 1995). Recent urban policy, however, particularly in the USA, appears to be more democratic, featuring community representation and some degree of local accountability.

Cultural issues

Culture has been an important component of the transformation of both the economies and the images of cities during the 1980s and 1990s. The cultures of cities have been affected in several important ways by their implication in the process of urban regeneration. 'Culture' includes not only cultural activities (entertainment, theatre, opera, ballet, music) associated with economic or social elites, but also those activities that constitute the 'everyday' or the 'ordinary', a more democratic notion of culture. These two meanings are often distinguished by the prefixes 'high' and 'popular'.

The cultural city

The development of new cultural facilities within cities has been led primarily by the need to appeal to wealthy external audiences and to complement urban regeneration developments such as prestigious hotels and convention centres. These facilities are typically very exclusive; they hold little appeal for the majority of the urban population either because of the activities that go on there or because of the expense involved. They may also be difficult to reach for populations with a heavy reliance on public transport, who once there may find the atmosphere unwelcoming. Further, it is likely that the programmes on offer will hold little appeal for certain groups, particularly ethnic or cultural minorities whose interests are usually very poorly provided for in international flagship developments (Bianchini *et al.* 1992).

The development of prestigious cultural facilities such as international concert halls is actually more likely to narrow the cultural profile of cities. This is so for two reasons. First, in most cases the development of prestigious cultural facilities goes hand-in-hand with attempts to attract major arts companies (e.g. ballet companies or orchestras), hosting cultural festivals of literature, drama, music, film, television and sport. A consequence of this is that attention within cities is shifted towards high culture of international standing and with international appeal. Less prestigious community-based cultural activities may suffer from cuts in funding, forcing them to restrict their activities or to close altogether. The brunt of these cuts falls on minority or community arts, especially of ethnic minorities (Lister 1991).

The use of culture within economic development has, on occasions, extended to the appropriation of formerly community-based and -led activities by local authorities, and subsequent incorporation into 'official' cultural programmes. The result may be that community groups lose control over activities that were formerly their own.

Excluded cultures and protest

The promotion of prestigious cultural facilities and activities of international appeal as part of programmes of urban regeneration has led to accusations from groups representing minority and community-based cultures that their contribution to the cultural life of cities is undervalued. Opposition from groups such as Workers' City, a group promoting the working-class history of Glasgow, to the European City of Culture celebrations mirrors a feeling that is widespread in cities undergoing similar 'make-overs'. These groups, which typically represent a diversity of cultures including working-class communities, ethnic minorities, women's organisations and gay and lesbian communities, have commonly felt that their culture has become devalued because it does not represent a 'marketable' commodity on the international circuit. This has prompted resistance by these groups to what they see as the shallow or 'facsimile' cultures promoted by urban regeneration. Such groups argue that they represent more 'genuine', 'organic' or deeply rooted cultures than those being promoted.

The means by which their opposition to the cultures promoted by urban regeneration is manifest are varied. Marches and demonstrations have long been a prominent form of urban protest. The history of the development of London's Docklands development from the early 1980s has been dogged by protest from community groups such as The Docklands Forum. Such demonstrations have included the 'People's Armada' in 1984 where a number of boats supporting community protests sailed past the Houses of Parliament (Rose 1992). Frequently opposition is expressed by less direct means and includes the articulation of community, history and a sense of belonging to a place threatened by development. Such protests have included publishing oppositional accounts of places often incorporating working-class social histories,

Plate 8.2 ***Graffiti protesting against the impacts of new development, Barcelona***

Source: Photograph, Matt Halstead

poster and arts campaigns, newsletters and petitioning local and national authorities (J.M. Jacobs 1992; Hall and Hubbard 1996).

The nature and organisation of oppositional groups have also been very diverse. Some are well organised and funded, and focus their intervention around specialist issues such as democratic planning and development, drawing upon the expertise of researchers and professionals, for example the community planning group Birmingham for People. Other groups employ the talents of artists to help articulate community feeling. A good example of this was the Docklands Community Poster Campaign involving the Art of Change group (Dunn and Leeson 1993). Other groups may be much more informal and spontaneous, forming to protest against specific developments. Some of the most frequently opposed aspects of urban regeneration developments include the physical, economic, cultural and psychological impacts of urban regeneration on communities, the lack of public consultation on development and the bypassing of existing planning legislation, and the lack of social and communal facilities in new developments.

Assessing the impact and influence of such groups is difficult, given their heterogeneity and the diversity of their aims and approaches. Their involvement with local authorities and developers may range from close consultation and involvement to outright hostility and confrontation. While it is dangerous to overgeneralise, it is probably true to say that the impact of these groups has been relatively marginal to the aims of local authorities and developers. Often the concessions granted to protesters have been token, or developers have appeared to take note of community issues largely to give the appearance of democracy, consultation and communal involvement in development (Hall and Hubbard 1996: 166).

Project idea

Collect census data to map levels of economic welfare in a city with which you are familiar from two recent Census periods, preferably about twenty years apart. Can you detect a distinctive geography of deprivation in your city? Is there any evidence of change to this geography over the period you have studied or has the pattern persisted? Can you find any explanations for the pattern you have observed?

Essay titles

- Discuss, with reference to examples, the assertion that social polarisation is inevitable in cities of the twenty-first century.
- Cultural development in cities has become closely alligned with economic development, effectively denying a place and a voice for 'alternative' cultural groups in cities. Discuss.

Further reading

Boger, J.C., and Webner, J.W. (eds) (1996) *Race, Poverty and American Cities*, Chapel Hill, NC: University of North Carolina Press. [Interesting essays exploring poverty in urban America with specific reference to its ethnic dimension, within the context of urban policy.]

Hamnett, C. (2003) *Unequal City: London in the Global Arena*, London: Routledge. [Detailed examination of economic change and its social consequences in London over the past forty years.]

Madanipor, A., Cars, G. and Allen, J. (eds) (1998) *Social Exclusion in European Cities: Processes, Experiences and Responses*, London: Jessica Allen. [Key collection of essays on European urban social exlcusion.]

Pacione, M. (ed.) (1997) *Britain's Cities: Geographies of Division in Urban Britain*, London: Routledge. [Excellent collection of essays covering a number of dimensions of inequality in British cities.]

Pacione, M. (2001) *Urban Geography: A Global Perspective*, London: Routledge (chapter 15). [Concise yet comprehensive review of key issues.]

Web resources

Neighbourhood Statistics (UK) http://neighbourhood.statistics.gov.uk

US Census Bureau www.census.gov

Joseph Rowntree Foundation www.jrf.org.uk

9 Sustainability and the city

Five key ideas

- **Cities are major contributors to global environmental problems.**
- **Environmental impacts within cities are socially mediated.**
- **Various models of sustainable urban form have been proposed.**
- **Western governments have argued that economic development and environmental sustainability are not incompatible (a stance referred to as ecological modernisation).**
- **Radical critics have argued that economic development is a fundamental cause of environmental unsustainability.**

Introduction

> Cities are the centre for the creation and redistribution of major environmental externalities. These are passed on unevenly, both within the city and outside.
>
> (Haughton and Hunter 1994: 52)

Traditionally, within mainstream urban geography, the focus has been overwhelmingly upon the human dimensions of the city. While geographers and others have looked at the physical geographies of the city (see e.g. in Michael Hough's *Cities and Natural Processes* (1995) and Ian Laurie's *Nature in Cities* (1979)), these have tended to have relatively little impact upon the practice of mainstream urban geography. This, however, is changing, and the issue is likely to represent one of the major foci of debate in urban geography in the coming years, and may well be so fundamental as to affect a major rethink of its practices and concerns. The late 1990s witnessed an explosion of debate among academics, politicians, pressure groups and in the media, about sustainability on global, regional and local scales. Much of this debate was focused on the city, which was identified as being a key building block in the path towards a more sustainable world. It is clear that cities have affected and will continue to fundamentally affect the development of the environment and that the environment should, if present concerns are taken seriously, affect the development of cities. For these reasons questions of sustainability and the interrelationship between cities and the environment are likely to emerge at the heart of urban geography in the future.

Having said all this, it would be wrong to suggest that questions of sustainability are purely environmental in their concerns. Rather, it is becoming apparent that many economic processes and social forms are unsustainable. For example, there were calls in late 1999 for certain deprived inner-city estates in the north of England to be demolished because they appeared to be in a state of terminal, irreversible decline. Putting aside the debate of whether or not this was the right course to take, it is clear that those estates had become, in their current state, unsustainable. Similarly, some have argued that outmigration from large cities in Europe and North America poses a threat to the future viability of some cities (Mohan 1999). The sustainability critique has, then, been applied both to the human and to the environmental dimensions of urbanisation. The concerns raised by this critique suggest not only that the economic and social viability of neighbourhoods, cities and their regions is in doubt, but also that their environmental viability is uncertain and that the processes of urbanisation as they are currently unfolding pose a series of threats to the global environment. It is likely in the future that geographers will examine, as well as issues of social justice and injustice generated by the processes of urbanisation, the questions of environmental justice and injustice they generate.

There are three key dimensions to the debate over the interrelationship between cities and the environment. These are outlined below. Many of the issues highlighted here are picked up in the subsequent discussion.

- *Cities as a threat to the environment* Cities are major contributors to global environmental problems including pollution, resource depletion and land take. While occupying a mere 2 per cent of the world's land surface, cities contain half of the world's population, which is increasing at a rate of 55 million people per year, consume three-quarters of the world's resources, and generate a majority of the world's waste and population (Blowers and Pain 1999: 249). This voraciousness is only likely to increase as urban growth rates increase, particularly in less developed countries. The environmental demands of city dwellers vary enormously. For example, city dwellers in developed countries typically generate up to twice as much waste per day as those in less developed countries (Haughton and Hunter 1994: 11). There are variations of a significant magnitude within, as well as between, developed and less developed countries. Whereas previously the consequences of these problems were primarily local, the scale of urbanisation and consequently of these problems ensures that their consequences are now global. For example, cities now typically draw on resources from all around the world, rather than from just their local region. Similarly, the pollution they generate is dispersed around the globe. The discovery of the depletion of the ozone layer starkly highlights the threat posed by cities and the processes fuelling their development.

- *The environment as a threat to cities* The environmental problems generated by cities are felt most severely within cities (Blowers and Pain 1999). Environmental problems such as pollution and its manifestations (for example, severe photochemical smogs) are both long established and increasingly apparent aspects of everyday life for many of the world's urban population. In addition, problems of land contamination from former industrial land use have imposed a severe constraint on urban development in many cities, and poses a very direct threat to individual health in some areas.
- *Social processes as mediators of environmental impacts and costs* In the same way as the impacts of, say, economic processes such as de-industrialisation are felt unevenly across different social groups (see Chapter 4), the environmental consequences and costs of urbanisation impact unevenly upon different social groups (Haughton and Hunter 1994). The environmental problems resulting from urbanisation tend to impact most severely upon the most vulnerable groups in urban society. These groups are less able to insulate themselves from the impacts of environmental problems. They tend to occupy marginal, sometimes contaminated, land, in shelter that may be unplanned or have only very basic amenities. The contrast between the risks borne by vulnerable groups from events such as floods or earthquakes, or from land contamination and sewerage discharge, for example, and the comparative lack of risks facing the wealthy in developing world cities, bears this out most clearly. To a lesser, but by no means insignificant, extent the same applies to the cities of the developed world.

The ecological footprints of cities

The sheer scale of the impacts of cities upon the environment is conveyed well by the concepts of the ecological footprints of cities (Blowers and Pain 1999; Massey 1999) and the global hinterlands of cities (Haughton and Hunter 1994). Maintaining contemporary cities requires that they draw upon and impact upon vast areas of land and water from beyond their own immediate geographical hinterlands. Cities draw resources, building materials, food, energy and so on from all over the world. They also disperse pollution and waste globally. These ecological connections stemming to and from cities may be visualised by the idea of the ecological footprint. Rees (1997: 305) defined the ecological footprints of cities as 'the total area of productive land and water required on a continuous basis to produce the resources consumed and to assimilate the wastes produced by that population, wherever on Earth the land is located'. For example, the ecological footprint of Vancouver is estimated as 180 times larger than the city's surface area, while that of London is put at 125 times larger than its surface area. The idea of the ecological footprint conveys the disproportionate impact of cities upon the environment. Clearly, if sustainable urban development is to be achieved, reducing the ecological footprints of cities is an imperative (Girardet 1996: 24–5; Blowers and Pain 1999).

Defining sustainability

Unsustainability implies that at some point in the future development will be compromised, or even threatened, as environmental capacity is reached or environmental limits are breached. This may involve a non-renewable resource depleting to the point of extinction or toxic emission to the point where it cannot be absorbed or dispersed without severe and irreparable damage to the environment. Unsustainability may be apparent at a variety of scales from the local to the global. At a local scale it is obvious that the environmental limits of many places have already been breached and that they have been rendered uninhabitable by over-exploitation or over-pollution as a result.

The question of what, and where, environmental limits and capacities are is a highly controversial and debated issue. Views on where these limits lie depend on whether one adopts a robust or a precautionary stance towards the environment's capacities (Mohan 1999). Furthermore, environmental capacities are far from fixed and absolute. They may expand as new technologies come on-stream, genetically modified organisms being one example. However, many apparent technological panaceas may compromise environmental capacities in other, initially unforeseen, ways.

The concept of sustainability, on the other hand, has been placed at the heart of many development paradigms in recent years. The idea of sustainable development implies that there are various paths which do not necessarily involve an inexorable march towards environmental overload but that can continue, in theory, indefinitely while remaining within, rather than breaching, environmental limits. The most commonly cited definition of sustainable development is the Brundtland definition put forward by World Commission on Environment and Development (1987): 'development that meets the needs of the present without compromising the ability of future generations to meet their own needs.'

Although useful, the difficulty with this definition is that needs are not absolute. Blowers and Pain (1999: 265) point out that 'what may be regarded as needs in the cities of the North would be luxuries in those of the South'. While offering a starting point to help understand the concept, the Brundtland definition is an idealistic and somewhat impractical definition of sustainable development.

Discussion topic

One of the key issues in the definition of sustainability is cross-cultural differences in the definition of needs, wants and luxuries. Can you think of a definition of 'needs' in the context of the developed world? What would you argue is a 'need' in this context? Indeed, is it possible to come to a consensus about this? Does this definition of 'needs' hold up when transferred to a developing world context? What are the implications of these cross-cultural differences for the definition and promotion of sustainable development?

Economic development and sustainability

There are two key aspects of the relationship between economic development and sustainable development worth considering here. First, there appears to be a conflict between short-term economic development and long-term environmental needs (Blowers 1997; Mohan 1999).

Governments in the developed world aim, primarily, to secure continued economic growth (Jacobs 1997). The benefits of this growth, for their people, are rising living standards that are defined in terms of increasing levels of personal consumption (Mohan 1999). Blowers (1997) has highlighted the incompatibility of individual lifestyles characterised by increasing levels of personal consumption and long-term collective interests in environmental sustainability. When we take into account that the aspirations of many governments of less developed countries are to obtain standards of living similar to those in the cities of the developed world, the problem becomes even more acute (Blowers and Pain 1999: 253–4).

Second, as earlier chapters have shown, one outcome of the current regime of capitalist development and flexible accumulation is heightened levels of social polarisation both within, and between, cities on a global scale. Such inequalities are inherently unsustainable socially, but also environmentally, and provide a significant barrier to achieving any level of sustainable development (Blowers and Pain 1999). The generation of poverty in cities is inextricably associated with the generation of environmental degradation.

Key dates and events

1972 Early publications – Meadows *et al. The Limits to Growth; Ecologist* 'Blueprint for survival'.

1976 Habitat I Forum (UN conference) – identified that city growth contributes to environmental problems. Advocated measures to slow or reverse urban growth.

1987 World Commission on Environment and Development – published *Our Common Future* (the Brundtland Report). Highlighted that inequalities between the developed and less developed world are a major barrier to global sustainable development.

1990 Publication of the *Green Paper on the Urban Environment* (European Commission) – advocated higher densities as a more sustainable urban form.

1992–94 Earth Summit, Rio de Janerio (UN conference) – led to the adoption of Agenda 21, the most comprehensive programme to date for

Key dates and events continued

	sustainable development. A key development was the involvement of local authorities in interpreting and implementing the issues covered by Agenda 21 (Local Agenda 21). This was signed in 1994.
1996	Habitat II (UN conference, 'the city summit') – an explicit exploration of the link between cities and sustainable development. Advocated organisational partnership and widespread participation as a key to achieving sustainable development. A major focus was on Local Agenda 21.

Sources: Haughton and Hunter (1994: 21–2); Breheny (1995: 83); Blowers and Pain (1999: 268)

Rather than seeking to restructure the regime of capitalist accumulation, the response of governments has been to argue, not surprisingly, that rising levels of economic development and personal consumption are not incompatible with sustainable development. This is a stance referred to as *ecological modernisation*, discussed in more detail below. Governments have sought to instigate measures that aim to secure sustainability without compromising levels of personal consumption. The stance of governments in the developed world has broadly been to encourage the 'greening' of capitalism without fundamentally affecting its operation. Measures adopted or proposed by governments have included internalising the costs of environmental externalities by incorporating them into the costs of industrial production. Although in many ways welcome, forcing companies to realise that production has an environmental cost, such measures are severely limited. It is possible with measures such as this for companies to continue to generate environmental externalities simply by paying for the 'right' to pollute (Haughton and Hunter 1994). Such measures may not even severely affect profit margins if these increased costs can be passed on to the consumer. Without global agreement and regulation, such measures are likely to be ineffective internationally. It is possible that companies may simply seek locations where the costs of environmental legislation on production are low or absent altogether. This demonstrates that a fundamental problem with much environmental legislation generated without a framework of global agreement is that it is only likely to result in the export of unsustainability to alternative locations. On a global scale such measures are likely to make limited contributions towards sustainability, a form of zero-sum growth.

Governments have also encouraged the 'greening' of businesses and this has become a significant aspect of corporate behaviour in the developed world. Again, however, such measures are likely to be very limited in their impact with the absence of global regulation of standards. While corporate take-up of green strategies remains voluntary, it is likely to be uneven. The types of measures that corporations have enacted have included the production of environmental statements or policies and the use of recyclable materials in packaging. However, critics have frequently questioned the substance of corporate environmental statements, particularly where companies have activities dispersed internationally. Critics have argued that corporate environmental statements frequently ignore the activities of branch plants in the developing world or subcontractors brought into the production process. They have referred to this as 'greenwash'.

There is a powerful North–South (developed world–less developed world) dimension to the promotion of sustainable development through ecological modernisation. The ecological modernisation stance is one advocated by developed world governments (Blowers and Pain 1999). It is a stance that has met with some resistance from developing world governments. These governments have argued that calls for the imposition of global environmental regulations and standards impose a barrier to development which unfairly penalises less developed nations, thus potentially preventing them from achieving levels of development currently enjoyed by more developed nations. They have argued that ecological modernisation is simply a way of perpetuating global inequality which, ironically, is likely only to perpetuate global unsustainability (Haughton and Hunter 1994). Developed nations have industrialised largely without the restriction of environmental legislation and accounting. However, many less developed nations regard it as unfair that, having already achieved high levels of development, developed nations are now seeking to deny less developed nations the environmental freedoms they themselves enjoyed.

Alternative paths towards sustainable economic development which while having only a limited impact globally, have had major impacts within specific localities include moves towards locally responsive and responsible enterprise. These include the development of locally owned small businesses, such as community food co-operatives (Haughton and Hunter 1994). Such businesses provide alternatives to expensive supermarkets or those located at out-of-town locations, inaccessible to deprived inner-city communities which are predominantly reliant upon impoverished public transport networks. A number of community-based credit unions and barter-based cashless exchange systems have also emerged, signalling the development of 'community economies' designed to help communities marginalised by dominant economic processes.

Reconciliation of economic development and sustainable development does not seem possible without some reconstruction of dominant understandings of what constitutes economic development. This would need to incorporate some aspect of long-term collective needs, both economic and environmental, rather than short-term individual demands (Haughton and Hunter 1994; Mohan 1999). Such a reconstruction is highly unlikely at anything beyond the most local of scales. This, and the community initiatives outlined above, suggests that the most appropriate scale at which alternative paradigms of economic development might be explored is the neighbourhood scale. There appears to be virtually no political will or understanding in developed countries to achieve this at any greater scale.

Urban size and form

There has been much debate around the question of whether there is an optimal city size in terms of sustainability. Debate generally associates increasing city size with unsustainability. It is true that increasing city size tends to be associated with increasing energy use per capita and that larger cities are more dependent upon external environmental resources. Finally, larger cities tend to be associated with greater environmental problems such as air and water pollution (Haughton and Hunter 1994).

Sustainability, however, involves a number of dimensions, and Michael Breheny, among others, has countered the arguments against increasing city size by highlighting ways in which large settlements might actually be more sustainable than smaller settlements. For example, urban decentralisation, contrary to accepted thinking, can potentially decrease commuting distances especially where employment decentralises to suburban locations (Breheny 1995: 84). Breheny also cites a study which suggests that the high population densities associated with large cities are actually characterised by relatively low car use. It is at low population densities that car travel tends to be highest, typically twice as high as at the highest population densities (ECOTEC 1993 cited in Breheny 1995: 84). Large cities also tend to have the most extensive and developed mass transit systems (Breheny 1995: 84). Others have noted that large cities have larger political and managerial infrastructures than smaller settlements, giving them greater political resources that can be mobilised, in theory at least, to manage and deal with environmental problems (Haughton and Hunter 1994: 75). Finally, Bairoch (1988) has noted that large cities have long been fertile cradles of innovation and the development of technologies. Such developments might conceivably contribute to enhanced sustainability in the future. Some of this innovation may well be borne out of the necessities for large cities to manage or solve their own environmental problems.

The question of optimal city size for sustainability is a complex one then, and one devoid of any easy answers. However, Haughton and Hunter (1994) are right to point out that of greater importance than urban size per se is the 'internal organisation' of cities, namely their form, processes and management. These issues are explored in subsequent sections.

A significant debate has also grown up around the relationship between urban form and sustainability. This follows widespread advocacy of various high-density compact city models by, for example, the European Union's *Green Paper on the Urban Environment* (1990), the Department of the Environment in the UK in their *Strategy for Sustainable Development* (1993), and a variety of planning groups and environmental pressure groups (Breheny 1995: 83). The basis of this advocacy is that the compact city, it is argued, offers a number of environmental benefits over more decentralised urban forms typical of the UK, USA and Australia. These supposed benefits include:

- Reduced fuel consumption through enhanced proximity and accessibility deriving from smaller city size and mixed-use development.
- A reconciliation of the needs for urban growth and rural protection (Breheny 1995).
- Accommodation of public transport provision.
- The potential for urban regeneration in dense mixed-use neighbourhoods (Blowers and Pain 1999).

Advocates of compact city models have also been quick to point out a number of environmental problems stemming from decentralised urban development. These have included:

- Extensive consumption of land.
- High rates of storm water pollution.

- High rates of water consumption through, for example, watering large garden plots.
- Increased fuel consumption through high car use and associated air pollution and health problems.
- High rates of energy consumption stemming from the poor thermal qualities of the single-storey and detached dwellings typical of decentralised urban forms.
- Low rates of waste recycling (Haughton and Hunter 1994: 85).
- A failure to accommodate extensive or efficient public transport provision.

Despite the appeal of the pro-compact city rhetoric, critics havey argued convincingly that there is little, if any, hard evidence to back it up and that, consequently, it is based largely around some questionable assertions. The most compelling critiques of the compact city model, and of the urban containment policies necessary to achieve it, are those advanced by Michael Breheny (1995). It is worth considering these in some detail.

Breheny advanced three main critiques of the advocacy of the compact city model. First is the fundamental point that compact city advocacy is informed by very little hard evidence, and rests primarily upon a set of false assumptions about both compact and decentralised urban forms. Breheny's review of the little evidence which does exist suggests that significant savings in fuel consumption following compact city development are highly unlikely to be realised. Even following an extreme policy of extensive re-urbanisation, rates of energy saving are only likely to be a modest 10 to 15 per cent.

> Would advocates of the compact city be so forceful if they knew that the likely gains from their proposals might be a modest 10 or 15 per cent energy saving achieved only after many years and unprecidentedly tough policies? Although proponents of the compact city have not generally specified expected levels of energy saving, it is to be assumed that they have in mind levels considerably higher than this.
>
> (Breheny, 1995: 95)

The wisdom of pursuing such extreme policies is further thrown into doubt when Breheny points out that similar levels of energy saving could be achieved through less painful methods such as incorporating improved technology into vehicle design and raising fuel costs (1995: 99). Breheny argues that compact city advocates who almost exclusively attack decentralised urbanisation have failed to take into account that other factors, such as household income and fuel prices, have a greater effect upon journey behaviour than urban form or size. In seeking to contain or reverse urban decentralisation, compact city models pursue the wrong quarry and are therefore unlikely to result in significant savings (Breheny 1995: 92).

Second, Breheny, among others, has argued that advocates of the, necessarily extreme, policy measures required to achieve compact city development have also failed to take into account the social costs of these measures. Given the prevalence of decentralisation as the dominant process of urban development in much of the developed world, achieving compact city development would involve nothing short of a fundamental reversal of this prevailing process. It is readily apparent that the policies required to achieve this are far from socially benign. It has been argued that the cramming and increased density that seems necessary is likely to result in an increase in overcrowding and homelessness in cities (Breheny and Hall 1996; Blowers and Pain 1999). Further, preservation of the rural environment may also come with unforeseen social consequences. Blowers and Pain argue that the lure of preserved rural environments would be in stark contrast to the increased population and building densities of the compact city. This is actually likely to increase flight to the suburbs and beyond for those who can afford it (see also Haughton and Hunter 1994; Blowers and Pain 1999). Taken together, Blowers and Pain argue, these problems suggest that compact city development is likely to heighten inequality within cities, itself a dimension perpetuating unsustainability.

In addition, Mohan (1999) has pointed out that imposing constraints on development, a further necessity of urban containment policies, in one locality may simply result in the relocation of environmentally harmful activities to areas without such restrictions. This is already apparent, although in a slightly different context, as one reason why multinational corporations relocate production globally is to exploit relatively lax environmental laws and regulations. This highlights the necessity for sustainability initiatives, of all kinds, to be situated within international systems of regulation if they are to be effective on a global scale.

Finally, Breheny argues that it is simply unrealistic to expect compact city models to be implemented given the current nature of urban development. He argues that, in the UK, for example, the forces of decentralisation are too strong, fuelled particularly by employment movement to the suburbs, and planning, political and national will too weak for any significant degree of urban containment to be realised (1995: 91). Urban containment policies operative in the UK in the recent past have brought only limited success, with development 'boundary jumping' a common problem. The widespread endorsement of city living necessary for the success of compact city models would involve a reversal of deeply held aspects of the cultures of many countries that heavily idealise the rural over the urban (Colls and Dodd 1986; Short 1992) and more tangibly goes against the wishes of many house builders and the market (Mohan 1999). Finally, the commitment to, and investment in, the existing urban realm is likely to create inherent inertia which may seriously mitigate against any significant change in urban form in the foreseeable future (Blowers and Pain 1999: 280). Neatly echoing this critique, Blowers and Pain (1999: 281) argue that compact city models for sustainable development are 'impractical, undesirable and unrealistic'.

Some commentators have also recognised actual and potential environmental benefits associated with decentralised urban forms. These benefits include the cooling effects of large gardens on high summer temperatures in hot countries, the potential for food production in large gardens, the potential for extensive rainwater reuse, space for solar panels (Haughton and Hunter, 1994: 89), and the potential for the recycling and composting of kitchen and organic waste. Recent studies in the UK have also highlighted the potentials of 'ordinary' gardens to foster biodiversity.

It appears that the most practical response to the apparently irreconcilable demands of unfolding urbanisation and sustainability is not to try and reverse decentralisation, but to focus growth around strong sub-centres, rather than allowing uniform low-density sprawl (Brotchie 1992). Gordon *et al.* (1991), for example, recognised that multi-centred urban areas actually tend to reduce commuting and hence motor fuel consumption per capita. In addition, strong urban sub-centres would allow the development of effective public transport networks within cities. Andrew Blowers (1993) has developed this theme in his model of the *MultipliCity*. In addition to developing strong public transport nodes, Blowers advocates urban infilling, some new settlement development and some development in rural areas as a practical path to more sustainable settlement forms within the context of current patterns of development. Crucial, Blowers argues, is provision for sustainability at the regional rather than simply the city scale.

Neighbourhoods and sustainability

It has been argued, echoing widespread criticisms of 'top-down' or universal solutions to all kinds of urban problems, that the key to developing sustainable urban forms lies in encouraging development at the neighbourhood level. This is certainly a more practical and realistic line of advocacy than that seeking to engineer massive structural change to urban forms. While neighbourhood advocacy should never reject the importance of higher levels of planning, it is clearly a recognition that overly centralised systems of government have failed to deliver economic and social regeneration to neighbourhoods, and seem unlikely to deliver environmental sustainability in the future. Neighbourhood-level planning thus has the potential to deliver solutions to a neighbourhood's own problems as well as providing the seed-bed for sustainability at greater scales (Carley 1999: 58).

The key to successful sustainable neighbourhood regeneration seems to lie in rejecting solutions or blueprints from outside in favour of those generated locally which build upon local expertise and knowledge and respect local conditions. Haughton and Hunter (1994: 114) refer to this as 'organic planning'. As we have seen, environmental sustainability is also dependent upon the operation of economic and social processes. Neighbourhood plans, therefore, should seek to integrate social and environmental aims if they are to be successful (Carley 1999). Experience has demonstrated that neighbourhood plans or visions are likely to be deliverable and sustainable only if they are based on widespread, extensive and lifelong participation. Participation, Carley (1999: 58) argues, should be regarded as a 'right of citizenship'.

Small-scale, neighbourhood-level planning seems to offer greater and more realistic potential for achieving sustainability than that dealing with more abstract and universal approaches. This echoes Haughton and Hunter's (1994) call for more holistic and flexible local solutions to problems of sustainability. It is worth remembering that what works in one neighbourhood is not necessarily likely to work in another.

Recycling of waste materials is also crucial to the development of more sustainable cities. Research has shown that neighbourhoods are the most effective level at which to develop recycling schemes and facilities, and in which to target information and education.

Case study I

Bedzed, sustainable living in South London

Large, sprawling metropolises, such as London, often face dual issues with regard to housing. They need to provide both affordable housing, for example, to key public sector workers such as nurses and teachers, and also to minimise the environmental impacts of development. One innovative scheme in the South London Borough of Sutton offers some clues to how these issues might be addressed in the future. The Beddington Zero Energy Development, or Bedzed for short, is a development of eighty-two houses, maisonettes and apartments, along with business spaces and a range of public open spaces and facilities, that offers a model for large-scale, affordable, sustainable housing.

There are three key elements to the Bedzed development that allow it to achieve its aims. These are: ecological design that minimises energy use; a high housing density, and a location that provides access to good public transport links. Each home uses sunlight for heating and electricity generation, and 80 per cent of the building materials used in construction are reclaimed timber and steel. The architect Bill Dunster said of the development: 'We are taking areas of relatively low-density suburban London and turning them into high-density live/work communities clustered around transport hubs.' The Bedzed development has been hailed as a possible solution to London's considerable housing crisis. For example, Sue Ellenby, Director of the London Housing Federation, said this scheme was 'the type of mixed tenure, mixed use, renewable-energy community we want to see more of'.

One significant hurdle that developments such as Bedzed face is the high building costs compared to traditional developments. This has made some developers wary of compromising profit margins. However, local authorities are increasingly being encouraged to account for the environmental and social benefits of schemes, as well as just the economic costs, in reaching decisions about developments.

Source: Minton (2001); see also www.bedzed.org.uk

Urban green spaces

Many plans for sustainable urban development involve calls to 'green the city'. This is a recognition that plants play important roles in 'moderating the impacts of human activities' in cities, such as absorbing emissions (Haughton and Hunter 1994: 118). Cities display enormous diversity in the range of their green spaces in terms of size and type. These spaces stretch from those of lone, individual plants through household gardens, verges and green corridors, parks, wastegrounds, sports fields and extensive planned civic spaces. It is apparent that the green spaces of cities, of whatever kind, are under severe pressure. They are being reduced in size and number as more profitable land uses acquire urban green spaces for development, and they are subject to increasingly artificial design (Haughton and Hunter 1994). However, 'greening' cities involves more than simply reducing or reversing the loss of urban green space. Successful strategies must pay attention also to the type of green spaces in cities, their size and their distribution. It is certainly not the case that more green space per se necessarily makes a greener city.

Hough (1995) has made the distinction between two types of green space typical of the city. He distinguishes between the 'pedigreed' landscape of formally planned, manicured civic spaces such as gardens and boulevards and the 'fortuitous', unplanned landscape of the city's forgotten and neglected spaces.

While the former spaces are revered as symbols of the highest civic virtues, part of the city's dominant landscape, Hough argues that in

Plate 9.1 ***The manicured civic landscape, Imperial Gardens, Cheltenham, UK***

Plate 9.2 ***The fortuitous landscape, an abandoned railway line, Cheltenham, UK***

many ways such landscapes are environmentally and ecologically problematic. He says that, first, such landscapes are universal – identical in contrasting urban settings worldwide. Consequently, these civic landscapes fail largely to reflect local cultures and ecologies. Second, he argues that they suppress diversity. The planting schemes of civic landscapes typically include only a very narrow range of species and are able to support very little wildlife. Further, weeds and alien plants are assiduously removed as part of the management schemes of civic spaces, rigorously reinforcing their narrow ecologies. These landscapes go against, rather than work with, natural processes. Finally, the management of these spaces is resource intensive, particularly with water, absorbing large quantities in plant and grass watering and in water features.

By contrast, Hough recognises a number of ecological advantages with the fortuitous landscape. These landscapes allow the development of diversity; being largely unmanaged, they consume few, if any, resources, and plants reflect local ecologies. Such landscapes, Hough argues, in contrast to the civic landscape, are a reflection of natural processes at work.

> If we make the not unreasonable assumption that diversity is ecologically and socially necessary to the health and quality of urban life, then we must question the values that have determined the image of nature in cities. A comparison between the plants and animals present in a landscaped residential front yard, or city park, reveals that the vacant lot has far greater floral and faunal diversity than the lawn or city park. Yet all efforts are directed towards nurturing the latter and suppressing the former. . . . The question that arises, therefore, is this: which are the derelict sites in the city requiring rehabilitation? Those fortuitous and often ecologically diverse landscapes representing urban natural forces at work, or the formlized landscapes created by design?
>
> (Hough 1995: 8–9)

A reaction to the environmental disadvantages stemming from artificially planned urban green spaces has led some to advocate the development of more fortuitous urban green spaces. One example of this has been the development of urban local nature reserves (LNRs) in the UK (Box and Barker 1998). These reserves must make special provision for studying and researching wildlife, or must preserve wildlife or other natural features of high value or special interest. LNRs are also able to satisfy the high value placed on wildlife and natural features by communities and are particularly important in providing study opportunities for schools. The number of LNRs in the UK has risen from twenty-four in 1970, to seventy-six in 1980, 236 in 1990 and 629 in 1997 (Box and Barker 1998: 360). LNRs are clearly an important part of the provision of green spaces for community amenity and education, and form important nodes in county or district environmental plans. The development of LNRs in the UK is clearly an important step in the reassessment of the value of different types of urban green spaces.

The size and distribution of urban green spaces is also important in terms of environmental sustainability. Large urban parks, such as Central Park in New York, are often celebrated and highly valued city landscapes. Despite their obvious beneficial psychological impact upon urban lives, urban parks may actually have negative impacts upon city ecologies. Such prominent urban parks are typically very artificially designed and sustain a narrow ecology like other civic landscapes. Furthermore, very large urban parks tend to break up functions and result in increased urban car use (Elkin *et al.* 1991; Haughton and Hunter 1994: 118). In addition, in being popular day-trip destinations, parks generate their own high traffic volumes.

Large parks in central city areas have also begun to attract their own severe social problems. Central Park in New York is regarded as highly unsafe at night and MacArthur Park near downtown Los Angeles has become a haven for high numbers of drug dealers and users. Studies in the UK have also revealed large parks and other green spaces to be sites characterised by high fears of crime and safety, especially for women. Problems include poor lighting in most parks and the presence of large numbers of potential hiding places for assailants (Burgess 1998). These problems clearly compromise the amenity value of large urban green spaces.

Large parks tend also to be widely dispersed in cities; many cities have only one such park and few other significant green spaces. Again this seems to bring ecological problems. The interconnectivity of urban green spaces is important in facilitating species migration. This has led many cities to develop green 'corridors', for example footpaths following disused railway lines.

Rather than developing a small number of widely dispersed spaces such as large parks, a more successful model of urban greening might involve spreading smaller parks throughout the city and developing neighbourhood, community or 'pocket' parks. It may be seen that balancing the needs of ecological diversity, amenity and safety is likely to emerge as a key challenge in urban green space planning and management. Clearly the vital role of parks in city life and culture cannot be ignored. However, an attitudinal shift is required that reveres green spaces in cities as much for their ecological as for their amenity values. A number of commentators have recognised the importance of achieving diversity in the provision of urban green spaces. Simply letting all urban green spaces 'go wild' is an unrealistic and undesirable option. It would seem more sensible and practical to advocate either the development of exclusively 'wild' spaces in cities, for example LNRs, alongside existing formal parks, or to allow wild spaces to develop within bigger formal green spaces (Haughton and Hunter 1994: 118; Hough 1995).

Urban green spaces are also able to contribute to local sustainability by being sites for the production of food. There has been growing interest recently in the potential of food growing in urban areas, especially among disadvantaged communities characterised by poor diets. A number of projects in the UK have combined food growing with nutrition and cookery classes, such as those in the Hartcliffe area of Bristol. There is a long tradition of food growing within Western cities on, for example, allotments (Crouch and Ward 1988), and formal and informal urban agriculture constitutes a major land use within developing world cities (Freeman 1991; Haughton and Hunter 1994).

The 'greening of the city' clearly has a number of contributions to make to the development of sustainable cities in a number of ways. However, it is crucial that any policy advocacy seeks to work with, rather than against, the internal biodiversity of cities, and that it seeks ways to reconcile the ecological need for green space in cities with those of amenity.

Discussion topic

One of the ways in which attitudes to nature in the city may be read and understood is through the definitions of 'pests'. Make a list of plants and animals that are commonly regarded as pests. What are the bases for these definitions? Do you regard the definition of pests in the city natural or as a social construction? Can you think of any implications of this for sustainable development?

Urban transport and sustainability

As previous sections have demonstrated, the question of sustainable transport is central to the sustainable development debate and is closely related to debates about economic development and urban size and form. The chief, almost the sole, focus of concern has been the increasing car use within and between cities. Increasing car ownership and use leads to a number of problems of urban sustainability. These include:

- Use of non-renewal resources (fuel, vehicle and road construction materials).
- Damage and loss of landscapes.
- Health problems, both respiratory problems and those involving traffic-related accidents.
- Emissions of greenhouse gases (Blowers and Pain 1999; Mohan 1999).
- Creation and exacerbation of social inequalities related to differential mobility.

Increased car use in cities is predominantly a reflection of an increase in average journey length (Haughton and Hunter 1994), the most rapid increases being for trips related to shopping and leisure activities as retail and leisure facilities have decentralised to suburban locations (Mohan 1999). However, at the same time unsustainable patterns of commuting to work have increased. Research from the United States has demonstrated that between 1980 and 1990 'drive-alone' commuting increased, accounting for 73.2 per cent of all commutes in 1990, compared to 64.4 per cent in 1980. This represented an increase of some twenty-two million commuters. This was significantly greater than the number of new workers over the same period. In the 1980s in the USA, therefore, all job growth was absorbed by drive-alone commuting (Pisarski 1992; Cervero 1995). The decentralisation of offices has also contributed to commuting-related problems of unsustainable transport development. New suburban office locations are designed to cater for car users, creating greater suburb-to-suburb commuting patterns. It is more difficult to provide viable public transport alternatives to such journeys than it is to suburb-to-centre commuting.

The shift towards flexible methods of production and working practices (see Chapter 4) also seems likely to increase car and motor vehicle use in cities. Flexible production is characterised by increased spatial fragmentation of production, small batch production and rapid response, 'just-in-time' production (Harvey 1989a). Such changes virtually preclude efficient and effective movement through mass freight transit methods, such as railways, that are relatively inflexible with regard to timetable and route. Flexible methods of production require large numbers of short journeys, typically in smaller vehicles from couriers or small hauliers (Mohan 1999). Similarly, the increase in flexible working practices (for example, flexible working hours) have begun to dissolve regular patterns of commuter movements in cities. It appears likely that as the working day becomes more flexible the traditional single mass movement of commuters from suburb to city centre will be replaced by a network of complex smaller commuter movements across the city at different times of the day. Again, serving such movements with viable public transport networks appears virtually impossible (Mohan 1999).

Some writers have recognised the potential of telecommuting and teleworking to alleviate high commuting volumes in cities in the future as telecommunications improve. The hope is that virtual commuting via computer terminals will replace actual commuting via car. While potential there clearly is, the reality is likely to be much less promising. There has been a significant increase in the numbers of people working from home, for all or part of their working week, in the past twenty years. However, this represents only a very small proportion of the total workforce. By 1990 in the USA only 3 per cent of the total workforce either wholly or partly 'teleworked' from home (Pisarski 1992). Barriers to widespread increases in teleworking include workers feeling that it led to feeling, or being, cut off from the social life of offices and from opportunities at work (Saloman 1984; Cervero 1995).

Prospects for reducing car use in cities in the future seem slim. This is both because reduced car use tends to penalise individuals where there is no equivalent public transport alternative and because facilities (i.e. retail and leisure) have tended to become dispersed following recent car-centred urban restructuring (Blowers and Pain 1999). Similarly, governments, at least in car-dependent Western societies, seem largely unable or unwilling to direct policy towards creating serious public transport alternatives to private car use (Mohan 1999). However, having said this it is possible to recognise a few innovative examples from around the world of cities that seem to have been successful in curbing the seemingly inevitable growth of car ownership and use (see case study below).

As well as the obvious environmental and health problems, car-dependent cities open up new dimensions of social inequality based around mobility. Increases in car ownership are socially uneven. At the same time decline in public transport, particularly following privatisation and deregulation, impact most severely upon the most disadvantaged groups in urban society. Furthermore, these groups tend to live in areas where the impacts of the problems created by the car are felt most severely (Blowers and Pain 1999). These problems include pollution and congestion as well as indirect car-generated problems such as the decline of neighbourhood facilities (for example, shops) in the face of competition from out-of-town facilities (Haughton and Hunter 1994).

As well as the widening social inequalities noted throughout this book we can now add those of the 'transport-rich' and the 'transport-poor'. There is evidence that the mobility-restricted transport-poor form a significant proportion of the urban population, with between 40 and 60 per cent of the urban population of many Western countries suffering reduced access to basic facilities (Engwicht 1992; Haughton and Hunter 1994).

> The car is probably the chief means by which environmental inequalities are created and sustained.
>
> (Reade 1997: 98)

> Like impoverished people throughout Britain, one of the[ir] major concerns is traffic. The lives of the poor are blighted by inequitable mobility: the capacity of the rest of the world to move past them, and their own incapacity to move away.
>
> (Monbiot 2000: 22)

Case study J

Curbing city car use: some international examples

A number of cities around the world have successfully managed to reduce car use. Examples include Singapore, Hong Kong, Curitiba in Brazil, Zurich, Toronto, and Portland in Oregon. The paths to achieving these reductions have varied, yet it is possible to derive a number of common principles from these examples. Four principles seem to be of particular significance:

- Planning for car reduction needs to take place at the city-wide scale.
- Cities must provide extensive public transport networks. These should attempt to be as efficient and flexible as the car as means of urban transport, and they should be integrated with land-use policy. Public transport infrastructure and vehicles should be of high quality and systems easily legible.
- Urban design and retail revitalisation should aim to make the public realm attractive and safe to encourage walking and cycling.
- Restrictions should be placed on car use, either through financial or physical access measures (such as reducing car-parking availability).

Perhaps the most developed example of a city designed to promote alternative travel is Curitiba in Brazil. Here scarce resources have been channelled into public transport rather than into car use. New urban developments have been encouraged along public transport routes so that the city actually grows and develops around its public transport networks. In addition, the city centre has been pedestrianised and many formerly busy roads have been converted into tree-lined walkways. The impacts of these measures have included: use of the public transport system by more than 1.3 million people per day; reductions in air pollution to among some of the lowest levels in Brazil, and a saving of 25 per cent on fuel consumption city-wide (Rabinovitch 1992).

Source: Newman (1996)

The politics of sustainability

The majority of developed world governments argue that sustainable development and economic growth are compatible. Indeed, they often argue that economic growth generates the capacity and resources with which unsustainability may be tackled. Their policies relating to sustainable development are built around incorporating ecological and environmental criteria into the production process, for example legislation controlling toxic emission (Blowers and Pain 1999). Such measures directly target environmental problems as manifest by, for example, a polluting factory. This stance sees unsustainability as the result, primarily, of a lack of attention by individuals or corporations to the environmental consequences of their activities. This may be addressed, those advocating an ecological modernisation stance would argue, by reminding these individuals and corporations of their responsibilities towards the environment through state legislation and regulation.

A number of critics of ecological modernisation adopt a far more radical stance. They argue, by contrast, that unsustainability, in both its social and environmental dimensions, is the inevitable outcome of current processes of development. They argue that this development depends upon the generation and perpetuation of economic and regional inequality and exploitation of the environment. Radicals see economic growth as the fundamental cause of unsustainability rather than as a path to its alleviation (Harvey 1996; Blowers and Pain 1999). This radical stance is typical of many environmental pressure groups.

These two stances appear, and in many ways are, incompatible. The former is located firmly within the institutions of the neo-liberal state, facilitating economic growth, the latter outside opposing it. The impact of the radical stance on the dominant ecological modernisation stance is clearly marginal. This is not to say that it is without influence, enjoying, as Harvey (1996) and Blowers and Pain (1999: 272) point out, 'striking successes both in drawing attention to problems and winning battles, especially at the local level'. This influence comes both through various forms of participation and, more typically, through opposition to development. However, it is clear that with such fundamental ideological divisions, sustainable development will continue to be the subject of vigorous political debate and struggle into the future, and vital in shaping all urban lives around the world.

Project idea

Collect a range of sustainable development policies operative in your area (e.g. your university or college, local authority, local or regional organisations, regional government, utility companies and private companies). In addition, try to collect information about specific events or initiatives aimed at promoting sustainable development. Can you 'map' the areas they identify for action (transport, recycling and so on) and the targets they attach to these? Can you identify any areas that are not addressed in these policies? Can you find evidence of these gaps being addressed by policies or initiatives from higher levels (national government or international agreements)? Do you feel these policies are co-ordinated? Based on this analysis and your wider reading, can you draw up a list of five priorities to ensure sustainable development within your local area?

Essay titles

- The compact city model of sustainable urban development is 'impractical, undesirable and unrealistic' (Blowers and Pain 1999: 281). To what extent do you agree with this statement?
- Do you believe that ecological modernisation can deliver global sustainable urban development?

Further reading

There are a number of excellent recently published books covering all the issues touched upon in this chapter, and many more besides, in far more detail than was possible here. Some of the best include:

Girardet, H. (2004) *Cities, People, Planet: Livable Cities for a Sustainable World*, Chichester: Wiley Academy. [Examines the contributions of architecture and design to sustainable urban development.]

Haughton, G. and Hunter, C. (1994) *Sustainable Cities*, London: Regional Studies Association. [An excellent, comprehensive overview of a number of key issues.]

Low, N., Gleeson, B., Elander, I. and Lidskog, R. (eds) (2000) *Consuming Cities: The Urban Environment in the Global Economy After Rio*, London: Routledge. [Topical collection examining cities as consumers of resources and the politics of sustainability.]

Shatterthwaite, D. (ed.) (1999) *The Earthscan Reader in Sustainable Cities*, London: Earthscan. [A valuable and wide-ranging collection of key readings.]

Williams, K., Burton, E. and Jenks, M. (eds) (2000) *Achieving Sustainable Urban Form*, London: Spon. [An extensive collection of essays primarily written from design, planning and architecture perspectives.]

Web resources

Sustainable Cities Network – www.hull.ac.uk/geog/research/html/suscity.html

10 Your urban geographies

It is important not to see urban geography as something that you study in higher education and then disregard before you move on to the next module or into a post-education career. Perhaps one of the things that attracted you to study urban geography in the first place was that you recognised that it was not abstract and theoretical but something rooted in the realities of everyday life for vast numbers of people. Hopefully as you have read through this book you will have recognised discussions of urban issues that are relevant to urban areas with which you are familiar. The ability to make the connections between what you see on the page and what you see on the street is one important way of making urban geography your own, something that clarifies your understanding of your world and that beyond your own immediate experience.

I said in the introductory chapter to this collection that it is difficult to engage with the major issues facing the world in the twenty-first century without engaging with urban geography in some way. What I would hope though is that in making these connections you adopt a critical perspective. It is important to use your own experiences and knowledge of the world to critically evaluate what you are told about urban geography. If you read Chapter 3 of this book it will be obvious to you that urban geography is not a static discipline. It has undergone a series of radical shifts throughout its history, at times rejecting previous perspectives. Urban geography today cannot offer you complete and total answers. It is important that you recognise this and think about the implications of instances where urban geography seems to say very little about an issue. Cities are changing more rapidly than urban geography. Often the subject is running to try to catch up. Recognising the failings of urban geography, as well as its strengths, is another important way that you can make urban geography your own.

Finally, perhaps the most significant way that you can make urban geography your own is to undertake a piece of original urban geographical research as part of an independent study or dissertation project. The range of urban research topics is huge, but hopefully this book has provided you with a variety of possible ideas. It is worth finishing by outlining a number of emergent debates in urban geography. You may want to choose one of these as a topic to research yourself. It is around these areas that some of the most interesting and exciting contributions are being made to the subject.

Emerging issues in urban geography

While it is always difficult to predict what will emerge as important areas of debate in the future, it is worth taking a broad view of emerging issues in urban geography. The issues outlined below are beginning to generate considerable interest among academics, policy-makers and various practitioners:

- Sustainable urban development
- New and changing forms of urban governance
- Social polarisation and social exclusion
- The impacts of technology on cities
- The impacts of the global economy on cities and the impacts of cities on the global economy

Further reading

Parsons, A.J. and Knight, P.G. (1995) *How to Do Your Dissertation in Geography and Related Disciplines*, London: Routledge. [An excellent subject-specific guide to all aspects of the dissertation process.]

Rodgers, A. and Viles, H. (eds) (2002) *The Student's Companion to Geography*, Oxford: Blackwell. [A number of good, accessible essays on both the subject and techniques of geography.]

References

Allen, J. (1988) 'Towards a post-industrial economy?' in Allen, J. and Massey, D. (eds) *The Economy in Question*, London: Sage.

Ambrose, P. (1994) *Urban Processes and Power*, London: Routledge.

Bailey, J.T. (1989) *Marketing Cities in the 1980s and Beyond: New Patterns, New Pressures, New Promises*, Cleveland, OH: American Economic Development Council.

Bairoch, P. (1988) *Cities and Economic Development: From the Dawn of History to the Present*, London: Mansell.

Barke, M. and Harrop, K. (1994) 'Selling the industrial town: identity, image and illusion' in Gold, J.R. and Ward, S.V. (eds) *Place Promotion: The Use of Publicity and Marketing to Sell Towns and Regions*, Chichester: Wiley.

Barnett, S. (1991) 'Selling us short', in Fisher, M. and Owen, U. (eds) *Whose Cities?*, Harmondsworth: Penguin.

Bassett, K. and Short, J.R. (1989) 'Development and diversity in urban geography', in Gregory, D. and Walford, R. (eds) *Horizons in Human Geography*, London: Macmillan.

Beazley, M., Loftman, P. and Nevin, B. (1995) 'Community resistance and mega-project development: an international perspective'. Paper presented to the British Sociology Association Annual Conference, University of Leicester.

Beckett, A. (1994) 'Take a walk on the safe side', *The Independent on Sunday Review*, (27 February): 10–12.

Bianchini, F. and Schwengal, H. (1991) 'Re-imagining the city', in Corner, S. and Harvey, J. (eds) *Enterprise and Heritage: Cross Currents of National Culture*, London: Routledge.

Bianchini, F., Dawson, J. and Evans, R. (1992) 'Flagship projects in urban regeneration', in Healey, P., Davoudi, S., O'Toole, M., Tavsanoglu, S. and Usher, D. (eds) *Rebuilding the City: Property-led Urban Regeneration*, London: Spon.

Bleeker, S. (1994) 'Towards the virtual corporation', *The Futurist*, (March–April): 11–14.

Blowers, A. (ed.) (1993) *Planning for a Sustainable Environment*, London: Earthscan.

Blowers, A. (1997) 'Society and sustainability', in Blowers, A. and Evans, B. (eds) *Town Planning into the 21st Century*, London: Routledge.

Blowers, A. and Pain, R. (1999) 'The unsustainable city', in Pile, S., Brook, C. and Mooney, G. *Unruly Cities? Order/Disorder*, London: Routledge/Open University.

Box, J. and Barker, G. (1998) 'Delivering sustainability through local nature reserves', *Town and Country Planning*, (December): 360–3.

Breheny, M. (1995) 'The compact city and transport energy consumption', *Transactions of the Institute of British Geographers*, 20(1): 81–101.

Breheny, M. and Hall, P. (1996) 'Four million households – where will they go?', *Town and Country Planning*, (February): 39–41.

Brotchie, J. (1992) 'The changing structure of cities', *Urban Futures*, (February): 13–23.

Burgess, J. (1998) 'Not worth taking the risk? Negotiating access to urban woodland', in Ainley, R. (ed.) *New Frontiers of Space, Bodies and Gender*, London: Routledge.

Byrne, D. (2001) *Social Exclusion*, Milton Keynes: Open University Press

Cameron, G.C. (1980) 'The economies of the conurbations', in Cameron, G.C. (ed.) *The Future of the British Conurbations*, London: Longman.

Carley, M. (1999) 'Neighbourhoods – building blocks of national sustainability', *Town and Country Planning*, (February): 58–60.

Carley, M. (2000) 'Urban partnerships, governance and the regeneration of Britain's cities', *International Planning Studies*, 5(3): 273–97.

Castells, M. (1977) *The Urban Question: A Marxist Approach*, London: Edward Arnold.

Castells, M. (1983) *The City and the Grassroots*, London: Edward Arnold.

Cervero, R. (1995) 'Changing live–work spatial relationships: implications for metropolitan structure and mobility', in Brotchie, J., Batty, M., Blakely, E., Hall, P. and Newton, P. (eds) *Cities in Competition: Productive and Competitive Cities for the 21st Century*, Melbourne: Longman Australia.

Champion, A.G. and Townsend, A.R. (1990) *Contemporary Britain: A Geographical Perspective*, London: Edward Arnold.

Christopherson, S. and Storper, M. (1986) 'The city as studio: the world as back lot: the impact of vertical disintegration on the location of the modern picture industry', *Environment and Planning D: Society and Space* 4(3): 305–20.

Cloke, P., Philo, C. and Sadler, D. (1991) *Approaching Human Geography: An Introduction to Contemporary Theoretical Debates*, London: Paul Chapman.

Cochrane, A. D. (2000) 'The social construction of urban policy', in Bridge, G. and Watson, S. (eds) *A Companion to the City*, Oxford: Blackwell.

Coleman, B.I. (ed.) (1973) *The Idea of the City in Nineteenth Century Britain*, London: Routledge & Kegan Paul.

Colls, R. and Dodd, P. (1986) *Englishness: Politics and Culture 1880–1920*, London: Croom Helm.

Commission to the European Communities (1990) *Green Paper on the Urban Environment*, Brussels: European Commission.

Cooke, P. (1990) 'Modern urban theory in question', *Transactions of the Institute of British Geographers*, (ns) 15(3): 331–43.

Cornwell, R. (1995) 'Joke City of the rust belt reborn in steel and glass', *The Independent*, (11 October): 13.

Cosgrove, D. (1989) 'Geography is everywhere: culture and symbolism in human landscapes', in Gregory, D. and Walford, R. (eds) *Horizons in Human Geography*, London: Macmillan.

Crilley, D. (1993) 'Architecture as advertising: constructing the image of redevelopment', in Kearns, G. and Philo, C. (eds) *Selling Places: The City as Cultural Capital, Past and Present*, Oxford: Pergamon.

Crouch, D. and Ward, C. (1988) *The Allotment: Its Landscape and Culture*, London: Faber.

Davis, M. (1990) *City of Quartz: Excavating the Future in Los Angeles*, London: Verso.

Davis, M. (2003) 'Fortress L.A.', in LeGares, R.T. and Stout, F. (eds) *The City Reader*, London: Routledge (3rd edn).

Department of the Environment (1993) *UK Strategy for Sustainable Development*, London: HMSO.

Diamond, J. (2001) 'Managing change or coping with conflict? Mapping the experience of a local regeneration partnership', *Local Economy*, 16: 272–85.

Domosh, M. (1989) 'A method for interpreting landscape: a case study of the New York World building', *Area*, 21(4): 347–55.

Domosh, M. (1992) 'Corporate cultures and the modern urban landscape of New York city', in Anderson, K. and Gale, F. (eds) *Inventing Places: Studies in Cultural Geography*, Melbourne: Longman/Wiley.

Duffy, H. (1990) 'The squeeze is on', *Financial Times* Section III: Relocation Survey, (26 April): 1.

Duncan, J. (1990) *The City as Text: The Politics of Landscape Interpretation in the Kandyan Kingdom*, Cambridge: Cambridge University Press.

Duncan, J. (1992) 'Elite landscapes as cultural (re)productions: the case of Shaughnessy Heights', in Anderson, K. and Gale, F. (eds) *Inventing Places: Studies in Cultural Geography*, Melbourne: Longman Cheshire.

Dunn, P. and Leeson, L. (1993) 'The art of change in Docklands', in Bird, J., Curtis, B., Putnam, T., Robertson, G. and Tickner, L. (eds) *Mapping the Futures: Local Cultures, Global Change*, London: Routledge.

Ecologist (1972) 'Blueprint for survival', *Ecologist*, 2,1.

ECOTEC (1993) *Reducing Transport Emissions Through Planning*, London: HMSO.

Elkin, T., McLaren, D. and Hillman, M. (1991) *Reviving the City: Towards Sustainable Urban Development*, London: Friends of the Earth.

Engwicht, D. (1992) *Towards an Eco-city: Calming the Traffic*, Sydney: Envirobook.

Evans, K., Taylor, I. and Fraser, P. (1996) *A Tale of Two Cities: Global Change, Local Feeling and Everyday Life in the North of England*, London: Routledge.

Eyles, J. (1987) 'Housing advertising as signs: locality creation and meaning-systems', *Geografiska Annaler*, 69B(2): 93–105.

Fincher, R. (1992) 'Urban geography in the 1990s', in Rogers, A., Viles, H. and Goudie, A. (eds) *The Student's Companion to Geography*, Oxford: Blackwell.

Fleming, T. (1999) 'Case studies for the creative industries', in Fleming, T. (ed.) *The Role of the Creative Industries in Local and Regional Development*, Manchester: Government Office for Yorkshire and Humberside/Forum on Creative Industries.

Freeman, D.B. (1991) *A City of Farmers: Informal Agriculture in the Open Spaces of Nairobi, Kenya* Montreal: McGill University Press.

Fretter, A.D. (1993) 'Place marketing: a local authority perspective', in Kearns, G. and Philo, C. (eds) *Selling Places: The City as Cultural Capital, Past and Present*, Oxford: Pergamon.

Garreau, J. (1991) *Edge City: Life on the New Frontier*, New York: Doubleday.

Girardet, H. (1996) *The Gaia Atlas of Cities' New Directions for Sustainable Urban Living*, London: Gaia Books.

Glass, R. (1989) *Clichés of Urban Doom and Other Essays*, Oxford: Blackwell.

Gold, J.R. and Gold, M. (1990) '"A place of delightful prospects": promotional imagery and the selling of suburbia', in Zonn, L. (ed.) *Place Images in Media*, Savage, MD: Rowman and Littlefield.

Gold, J.R. and Gold, M. (1994) '"Home at last": building societies, home ownership and the imagery of English suburban promotion in the interwar years', in Gold, J.R. and Ward, S.V. (eds) *Place Promotion: The Use of Publicity and Marketing to Sell Towns and Regions*, Chichester: Wiley.

Gold, J.R. and Ward, S.V. (eds) (1994) *Place Promotion: The Use of Publicity and Marketing to Sell Towns and Regions*, Chichester: Wiley.

Goodwin, M. (1992) 'The changing local state', in Cloke, P. (ed.) *Policy and Change in Thatcher's Britain*, Oxford: Pergamon.

Gordon, P., Richardson, H.W. and Jun, M.J. (1991) 'The commuting paradox: evidence from the top twenty', *Journal of the American Planning Association*, 57(4): 416–20.

Gottdiener, M. (1986) 'Recapturing the center: a semiotic analysis of shopping malls', in Gottdiener, M. and Lagopoulos, A. (eds) *The City and the Sign: An Introduction to Urban Semiotics*, New York: Columbia University Press.

Graham, S. and Marvin, S. (1996) *Telecommunications and the City: Electronic Spaces, Urban Places*, London: Routledge.

Greed, C. (1993) *Introducing Town Planning*, Harlow: Longman.

Hall, P. (1995) 'Urban stress, creative tension', *Independent*, (21 February): 15.

Hall, T. (2004) 'Art and urban regeneration', in Blunt, A., Gruffudd, P., May, J., Ogborn, M. and Pinder, D. (eds) *Cultural Geography in Practice*, London: Arnold.
Hall, T. and Hubbard, P. (1996) 'The entrepreneurial city: new urban politics, new urban geographies?', *Progress in Human Geography*, 20(2): 153–74.
Hall, T. and Hubbard, P. (1998) *The Entrepreneurial City: Geographies of Politics, Regime and Representation*, Chichester: Wiley.
Hambleton, R. (1991) 'American dreams, urban realities', *The Planner*, 77(23): 6–9.
Hambleton, R. (1995) 'The Clinton policy for cities: a transatlantic assessment', *Planning Practice and Research*, 10(3/4): 359–77.
Hamnett, C. (1991) 'The blind men and the elephant: the explanation of gentrification', *Transactions of the Institute of British Geographers*, (ns) 16(2): 173–89.
Hamnett, C. (1995) 'Controlling space: global cities', in Allen, J. and Hamnett, C. (eds) *A Shrinking World: Global Unevenness and Inequality*, Oxford: Oxford University Press.
Hartshorn, T. and Muller, P.O. (1989) 'Suburban downtowns and the transformation of metropolitan Atlanta's business landscape', *Urban Geography* 10: 375–95.
Harvey, D. (1973) *Social Justice and the City*, London: Edward Arnold.
Harvey, (1979, 1989) 'Monument and myth', *Annals of the Association of American Geographers*, reprinted in Harvey, D. *The Urban Experience*, Oxford: Blackwell.
Harvey, D. (1988) 'Voodoo cities' *New Statesman and Society* (30 September): 33–5.
Harvey, D. (1989a) *The Condition of Postmodernity* Oxford: Blackwell.
Harvey, D. (1989b) *The Urban Experience*, Oxford: Blackwell.
Harvey, D. (1996) 'The environment of justice', in Merrifield, A. and Swyngedouw, E. (eds) *The Urbanization of Injustice*, London: Lawrence & Wishart.
Haughton, G. and Hunter, C. (1994) *Sustainable Cities*, London: Regional Studies Association.
Healey, M. and Ilbery, B.W. (1990) *Location and Change: Perspectives on Economic Geography*, Oxford: Oxford University Press.
Healey, P., Magalhaes, C. de, Madanipour, A. and Pendlebury, J. (2002) *Shaping City Centre Futures: Conservation, Regeneration and Institutional Capacity*, Newcastle upon Tyne: Centre for Research in European Urban Environments, University of Newcastle.
Hewison, R. (1987) *The Heritage Industry: Britain in a Climate of Decline*, London: Methuen.
Holcomb, B. (1990) *Purveying Places Past and Present*, New Brunswick, NJ: Urban Policy Research Working Paper no. 17.
Holcomb, B. (1993) 'Revisioning place: de- and re-constructing the image of the

industrial city', in Kearns, G. and Philo, C. (eds) *Selling Places: The City as Cultural Capital, Past and Present*, Oxford: Pergamon.

Holcomb, B. (1994) 'City make-overs: marketing the post-industrial, North American city', in Gold, J.R. and Ward, S.V. (eds) *Place Promotion: The Use of Publicity and Marketing to Sell Towns and Regions*, Chichester: Wiley.

Hough, M. (1995) *Cities and Natural Processes*, London: Routledge.

Howden, D. (2005) 'Wanted: a home for the "herd of white elephants" left by the Athens Olympics', *Independent* (01 April): 27.

Hubbard, P.J. (1996) 'Re-imaging the city: the transformation of Birmingham's urban landscape', *Geography*, 81(1): 26–36.

Hudson, R. (1989) 'Yacht havens in a sea of despair', *Times Higher Education Supplement* (20 January): 18.

Hudson, R. and Williams, A. (1986) *The United Kingdom*, London: Harper & Row.

Imrie, R. and Raco, M. (eds) (2003) *Urban Renaissance? New Labour, Community and Urban Policy*, Bristol: Policy Press.

Imrie, R. and Thomas, H. (1993) 'The limits of property-led regeneration' *Environment and Planning C: Government and Policy*, 11(1): 87–102.

Imrie, R., Thomas, H. and Marshall, T. (1995) 'Business organisation, local dependence and the politics of urban renewal, in Britain' *Urban Studies*, 32(1): 31–47.

Jackson, P. and Holbrook, B. (1995) 'Multiple meanings: shopping and the cultural politics of identity', *Environment and Planning A*, 27(12): 1913–30.

Jacobs, J.M. (1992) 'Cultures of the past and urban transformation: the Spitalfields market redevelopment in East London', in Anderson, K. and Gale, F. (eds) *Inventing Places: Studies in Cultural Geography*, Melbourne: Longman.

Jacobs, M. (ed.) (1997) *Greening the Millennium: The New Politics of the Environment*, Oxford: Blackwell.

Jameson, F. (1992) *Postmodernism, or the Cultural Logic of Late Capitalism*, London: Verso.

Jencks, C. (1984) *The Language of Postmodern Architecture*, London: Academy Editions.

Johnston, R.J., Taylor, P.J. and Watts, M.L. (eds) (1995) *Geographies of Global Change: Remapping the World in the Late Twentieth Century*, Oxford: Blackwell.

Keil, R. (1994) 'Global sprawl: urban form after Fordism?', *Environment and Planning D: Society and Space*, 12(2): 131–6.

Kelly, A. (2004) 'Bermondsey takes the biscuit', *Guardian*, (20 October): 4.

Knopp, L. (1987) 'Social theory, social movements and public policy: recent accomplishments of the gay and lesbian movements in Minneapolis, Minnesota', *International Journal of Urban and Regional Research*, 11(2): 243–61.

Knowles, R. and Wareing, J. (1976) *Economic and Social Geography Made Simple*, London: Heinemann.

Knox, P.L. (1991) 'The restless urban landscape: economic and socio-cultural change and the transformation of Washington D.C.', *Annals of the Association of American Geographers*, 81(2): 181–209.

Knox, P.L. (1992a) 'The packaged landscapes of postsuburban America', in Whitehand, J.W.R. and Larkham, P.J. (eds) *Urban Landscapes: International Perspectives*, London: Routledge.

Knox, P.L. (1992b) 'Suburbia by stealth', *Geographical Magazine* (August): 26–9.

Knox, P.L. (ed.) (1993) *The Restless Urban Landscape*, Englewood Cliffs, NJ: Prentice-Hall.

Knox, P.L. (1995) 'World cities and the organisation of global space', in Johnston, R.J., Taylor, P.J. and Watts, M.J. (eds) *Geographies of Global Change: Remapping the World in the Late Twentieth Century*, Oxford: Blackwell.

Knox, P.L. and Agnew, J. (1994) *The Geography of the World Economy*, London: Edward Arnold (2nd edn).

Lauria, M. and Knopp, L. (1985) 'Towards an analysis of the role of gay communities in the urban renaissance', *Urban Geography* 6(2): 152–69.

Laurie, I.C. (ed.) (1979) *Nature in Cities: The Natural Environment in the Design and Development of Urban Green Space*, Chichester: Wiley.

Lewis, P. (1983) 'The galactic metropolis', in Platt, R.H. and Macinko, G. (eds) *Beyond the Urban Fringe*, Minneapolis, MN: University of Minnesota Press.

Ley, D. (1983) *A Social Geography of the City*, New York: Harper & Row.

Ley, D. (1989) 'Modernism, postmodernism and the struggle for place', in Agnew, J. and Duncan, J. (eds) *The Power of Place*, London: Unwin Hyman.

Ley, D. and Mills, C. (1993) 'Can there be a postmodernism of resistance in the urban landscape?', in Knox, P. (ed.) *The Restless Urban Landscape*, Englewood Cliffs, NJ: Prentice-Hall.

Ley, D. and Olds, K. (1988) 'Landscape as spectacle: world's fairs and the culture of heroic consumption', *Environment and Planning D: Society and Space*, 6(2): 191–212.

Leyshon, A. (1995) 'Annihilating space? The speed-up of communications', in Allen, J. and Hamnett, C. (eds) *A Shrinking World? Global Unevenness and Inequality*, Oxford: Oxford University Press.

Leyshon, A. and Thrift, N. (1994) 'Access to financial services and financial infrastructure withdrawal: problems and policies', *Area*, 26, 3: 268–75.

Leyshon, A., Thrift, N. and Daniels, P. (1990) 'The operational development and spatial expansion of large commercial development firms', in Healey, P. and Nao, R. (eds) *Land and Property Development in a Changing Context*, Brookfield, VT: Gower.

Lister, D. (1991) 'The transformation of a city: Birmingham', in Fisher, M. and Owen, U. (eds) *Whose Cities?*, Harmondsworth: Penguin.

Loftman, P. (1990) *A Tale of Two Cities: Birmingham the Convention and Unequal City. The International Convention Centre and Disadvantaged Groups*, Birmingham: Birmingham Polytechnic, Faculty of the Built Environment.

Loftman, P. and Nevin, B. (1992) *Urban Regeneration and Social Equity: A Case Study of Birmingham 1986–1992*, Birmingham: University of Central England, School of Planning.

Loftman, P. and Nevin, B. (1994) 'Prestige project developments: economic renaissance or economic myth?', *Local Economy*, 11(4): 307–25.

Malecki, E. (1991) *Technology and Economic Development: The Dynamics of Local, Regional and National Change*, London: Longman.

Markusen, A. (1983) 'High tech jobs, markets and economic development prospects', *Built Environment*, 9(1): 18–28.

Massey, D. (1984) *Spatial Divisions of Labour*, London: Macmillan.

Massey, D. (1999) 'Cities in the world', in Massey, D., Allen, J. and Pile, S. (eds) *City Worlds*, London: Routledge/Open University.

Massey, D. and Meegan, R. (1982) *The Anatomy of Job Loss: The How, Why and Where of Employment Decline*, London: Macmillan.

Mayne, A. (1993) *The Imagined Slum: Newspaper Representation in Three Cities*, Leicester: Leicester University Press.

McCracken, G. (1988) *Culture and Consumption: New Approaches to the Symbolic Character of Consumer Goods and Activities*, Bloomington, IN: Indiana University Press.

Meadows, D.H., Meadows, D.L., Randers, J. and Behrens, W.W. III (1972) *The Limits to Growth: A Report for the Club of Rome's Project on the Predicament of Mankind*, Boston, MA: MIT Press.

Miles, M. (1997) *Art, Space and the City*, London: Routledge.

Miller, D., Jackson, P., Thrift, N., Holbrook, B. and Rowlands, B. (1998) *Shopping, Place and Identity*, London: Routledge.

Minton, A. (2001) 'Rising sun', *Guardian*, (19 September): 5.

Mohan, J. (1999) *A United Kingdom? Economic, Social and Political Geographies*, London: Arnold.

Monbiot, G. (2000) 'Complainers who are facing ruin', *Guardian*, (10 February): 22.

Montgomery, A. (1999) 'Proof of the pudding', *Hotline*, (July): 32–5.

Montgomery, J. (1994) 'The evening economy of cities', *Town and Country Planning*, (November): 302–7.

Moudon, A.V. (1992) 'The evolution of twentieth century residential forms: an American case study', in Whitehand, J.W.R. and Larkham, P.J. (eds) *Urban Landscapes: International Perspectives*, London: Routledge.

Newman, P. (1996) 'Transport: reducing automobile dependence', in Shatterwaite, D. (ed) *The Earthscan Reader in Sustainable Cities*, London: Earthscan.

O'Connor, J. and Wynne, D. (eds) (1996) *From the Margins to the Centre*, Aldershot: Arena.

Oatley, N. (1993) 'Realising the potential for urban policy: the case of Bristol Development Corporation', in Imrie, R. and Thomas, H. (eds) (1993) *British Urban Policy and the Urban Development Corporations*, London: Paul Chapman.

Oatley, N. (ed.) (1998) *Cities, Economic Competition and Urban Policy*, London: Paul Chapman.

Pacione, M. (2001) *Urban Geography: A Global Perspective*, London: Routledge.

Page, S. (1995) *Urban Tourism*, London: Routledge.

Pendlebury, J. (2002) 'Conservation and regeneration: complementary or conflicting processes. The case of Grainger Town, Newcastle Upon Tyne', *Planning Practice and Research* 17(2): 145–58.

Pisarski, A.S. (1992) *New Perspectives on Commuting*, Washington, DC: Federal Highways Administration.

Raban, J. (1986) *Old Glory*, London: Picador.

Rabinovitch, J. (1992) 'Curitiba: towards sustainable urban development', *Environment and Urbanisation*, 4, 2: 62–73.

Rappaport, A. (1990) *The Meaning of the Built Environment*, Tucson, AZ: University of Arizona Press (2nd edn).

Reade, E.J. (1997) 'Planning in the future or planning of the future?', in Blowers, A. and Evans, B. (eds) *Town Planning into the 21st Century*, London: Routledge.

Rees, W. (1997) 'Is "sustainable city" an oxymoron?', *Local Environment*, 2(3): 303–10.

Reid, L. and Smith, N. (1993) 'John Wayne meets Donald Trump: The Lower East Side as wild wild west', in Kearns, G. and Philo, C. (eds) *Selling Places: The City as Cultural Capital, Past and Present*, Oxford: Pergamon.

Relph, E. (1976) *Place and Placelessness*, London: Pion.

Rex, J. and Moore, R. (1967) *Race, Community and Conflict: A Study of Sparkbrook*, Harmondsworth: Penguin.

Rex, J. and Tomlinson, S. (1979) *Colonial Immigrants in a British City*, London: Routledge & Kegan Paul.

Roberts, P. (2000) 'The evolution, definition and purpose of urban regeneration', in Roberts, P. and Sykes, H. (eds) *Urban Regeneration: A Handbook*, London: Sage.

Roberts, P. and Sykes, J. (eds) (2000) *Urban Regeneration: A Handbook*, London: Sage.

Rose, D. (1989) 'A feminist perspective of employment restructuring and gentrification: the case of Montreal', in Wolch, J. and Dear, M. (eds) *The Power of Geography*, London: Unwin Hyman.

Rose, G. (1992) 'Local resistance to the LDDC: community attitudes and action', in Ogden, P. (ed.) *London Docklands: The Challenge of Development*, Cambridge: Cambridge University Press.

Rowley, G. (1994) 'The Cardiff Bay Development Corporation: urban regeneration, local economy and community', *Geoforum* 25, 3: 265–84.

Ryan, K.B. (1990) 'The "official" image of Australia', in Zonn, L. (ed.) *Place Images in Media*, Savage, MD: Rowman and Littlefield.

Saloman, I. (1984) 'Telecommuting: promises and reality', *Transport, Reviews* 4(1): 103–13.

Samuel, R. (1994) *Theatres of Memory: Past and Present in Contemporary Culture*, London: Verso.

Sassen, S. (1991) *The Global City: New York, London, Tokyo*, Princeton, NJ: Princeton University Press.

Sassen, S. (1994) *Cities in a World Economy*, Thousand Oaks, CA: Pine Forge Press.

Saunders, P. (1990) *A Nation of Homeowners*, London: Unwin Hyman.

Savage, M. and Warde, A. (1993) *Urban Sociology, Capitalism and Modernity*, London: Macmillan/British Sociological Association.

Scott, A. (1988) *Metropolis: From the Division of Labour to Urban Form*, Berkeley, CA: University of California Press.

Scott, K. (2004) 'As the health and wealth gaps widen, Glasgow rebrands itself as a city of style', *Guardian* (10 March): 9.

Selman, P. (1996) *Local Sustainability, Managing and Planning Ecologically Sound Places*, London: Sage.

Shiner, P. (1995) 'Urban regeneration: making less seem more', *Guardian (G2)* (11 January): 39.

Short, B. (ed.) (1992) *The English Rural Community: Image and Analysis*, Cambridge: Cambridge University Press.

Short, J.R. (1984) *An Introduction to Urban Geography*, London: Routledge & Kegan Paul.

Short, J.R. (1989) 'Yuppies, yuffies and the new urban order', *Transactions of the Institute of British Geographers*, (ns) 14, 2: 173–88.

Short, J.R., Benton, L.M., Luce, W.B. and Walton, J. (1993) 'Reconstructing the image of the industrial city', *Annals of the Association of American Geographers*, 83, 2: 207–24.

Smith, D.J. (1977) *Racial Disadvantage in Britain*, Harmondsworth: Penguin.

Smith, N. and Williams, P. (eds) (1986) *Gentrification of the City*, Boston, MA: Unwin Hyman.

Soja, E.W. (1989) *Postmodern Geographies: The Reassertion of Space in Critical Social Theory*, London: Verso.

Soja, E.W. (1995) 'Postmodern urbanization: the six restructurings of Los Angeles', in Watson, S. and Gibson, K. (eds) *Postmodern Cities and Spaces*, Oxford: Blackwell.

Soja, E.W. (1996) *Thirdspace: Journeys to Los Angeles and Other Real and Imagined Places*, Oxford: Blackwell.

Speake, J. and Fox, V. (2002) *Regenerating City Centres*, Sheffield: Geographical Association.

Spencer, K., Taylor, A., Smith, B., Mawson, J., Flynn, N. and Batley, R. (1986) *Crisis in the Industrial Heartland: A Study of the West Midlands*, Oxford: Clarendon Press.

Stimson, R.J. (1995) 'Processes of globalisation, economic restructuring and the emergence of a new space economy of cities and regions in Australia', in Brotchie, J., Batty, M., Blakely, E., Hall, P. and Newton, P. (eds) *Cities in Competition: Productive and Competitive Cities for the Twenty-First Century*, Melbourne: Longman Australia.

Stock, B. (1986) 'Texts, readers and enacted narratives', *Visible Language*, 20: 294–301.

Strassman, W.P. (1988) 'The United States', in Strassman, W.P. and Wells, J. (eds) *The Global Construction Industry*, London: Unwin Hyman.

Sudjic, D. (1993) *The 100 Mile City*, London: Flamingo.

Taylor, P. and Bain, P. (2003) *Call Centres in Scotland and Outsourced Competition from India*, Stirling: Scotecon.

Teather, E.K. (1991) 'Visions and realities: images of early postwar Australia', *Transactions of the Institute of British Geographers*, (ns) 16, 4: 470–83.

The Economist (1995) 'From screwdrivers to science'. (18 March): 32–4.

Thomas, H. (1994) 'The local press and urban renewal: a South Wales case study', *International Journal of Urban and Regional Studies*, 18(2): 315–33.

Thrift, N. (1987) 'The fixers: the urban geography of international commercial capital', in Henderson, J. and Castells, M. (eds) *Global Restructuring and Territorial Development*, London: Sage.

Turok, I. (1992) 'Property-led regeneration: panacea or placebo?', *Environment and Planning A*, 24(3): 361–79.

Turok, I. (1993) 'Inward investment and local linkages: how deeply embedded is Silicon Glen?', *Regional Studies*, 27(5): 401–17.

Ward, S.V. (1988) 'Promoting holiday resorts: a review of early history to 1921', *Planning History*, 10(1): 7–11.

Ward, S.V. (1990) 'Local industrial promotion and development policies 1899–1940', *Local Economy*, 5(2): 100–18.

Ward, S.V. (1994) 'Time and place: key themes in place promotion in the USA, Canada and Britain since 1870', in Gold, J.R. and Ward, S.V. (eds) *Place Promotion: The Use of Publicity and Marketing to Sell Towns and Regions*, Chichester: Wiley.

Watson, S. (1991) 'Gilding the smokestacks: the new symbolic representations of deindustrialised regions', *Environment and Planning D: Society and Space*, 9(1): 59–71.

Whitehand, J.W.R. (1990) 'Makers of the residential townscape: conflict and change in outer London', *Transactions of the Institute of British Geographers*, (ns) 15(1): 87–101.

Whitehand, J.W.R. (1994) 'Development cycles and urban landscapes', *Geography*, 79(1): 3–17.

Whitehand, J.W.R. and Larkham, P.J. (eds) (1992) *Urban Landscapes: International Perspectives*, London: Routledge.

Whitehand, J.W.R., Larkham, P.J. and Jones, A.N. (1992) 'The changing suburban landscape in post-war England', in Whitehand, J.W.R. and Larkham, P.J. (eds) *Urban Landscapes: International Perspectives*, London: Routledge.

Wolman, H.L., Ford, C.C. III and Hill, E. (1994) 'Evaluating the success of urban success stories', *Urban Studies*, 31(6): 835–50.

World Commission on Environment and Development (1987) *Our Common Future*, Oxford: Oxford University Press.

Wynn Davies, P. (1992) 'Development losses of £67m "a scandal"', *Independent* (13 July): 2.

Young, C. and Lever, J. (1997) 'Place promotion, economic location and the consumption of city image', *Tijdschrift voor Economische en Sociale Geografie*, 88(4): 332–41.

Zukin, S. (1988) *Loft Living: Culture and Capital in Urban Change*, London: Radius.

Index

Note: page numbers in *italics* denote references to illustrations and tables.

www.study-guides.com
www.study-guides.com
www.study-guides.com
www.study-guides.com